Cambridge University Press India Pvt. Ltd.
Under the Foundation Books imprint
Cambridge House, 438/1/4 Ansari Road, Daryaganj, **New Delhi** 110002

Cambridge University Press India Pvt. Ltd.
C-22, C-Block, Brigade M.M., K.R. Road, Jayanagar, **Bangalore** 560 070
Plot No. 80, Service Industries, Shirvane, Sector-1, Nerul, **Navi Mumbai** 400 706
10, Raja Subodh Mullick Square, 2nd Floor, **Kolkata** 700 013
21/1 (New No. 49), 1st Floor, Model School Road,
Thousand Lights, **Chennai** 600 006
House No. 3-5-874/6/4, (Near Apollo Hospital), Hyderguda, **Hyderabad** 500 029

Jointly published by Cambridge University Press India Pvt. Ltd. and International
Development Research Centre.

International Development Research Centre
PO Box 8500
Ottawa, ON KIG 3H9
Canada
www.idrc.ca/info@idrc.ca
ISBN (e-book) 978-1-55250-368-3

Typeset by Amrit Graphics, Shahdara, Delhi 110 032

ISBN 978-81-7596-521-8

Published by Manas Saikia for Cambridge University Press India Pvt. Ltd. and
printed at Sanat Printers, Kundli.

An Allegory of Prudence
By Titian and workshop

"Ex praeterito praesens prudenter agit
ni futura actione deturbet."

"From experience of the past
the present acts prudently
lest it despoil future action."

Development at Risk

Ignoring the Past

Joseph H. Hulse

International Development Research Centre

Ottawa • Cairo • Dakar • Montevideo • Nairobi • New Delhi • Singapore

ƒOUNDATION
BOOKS

Delhi • Bangalore • Mumbai • Kolkata • Chennai • Hyderabad

Dedication

This book is dedicated to the memory of Professor Y. Nayudamma, a dear friend, an exceptional Indian scientist and a former member of IDRC's Board of Governors.

His life was a testament to his often-stated conviction that "Science serves no useful purpose if it does not serve to satisfy the needs of humanity, particularly those who are in greatest need".

Contents

Foreword

Professor M.G.K. Menon

When my good friend, Dr. J.H. – Joe to me – Hulse spoke to me over the telephone, requesting a Foreword for a book that he was writing on "Sustainable Development at Risk: Ignoring the Past", I readily agreed for many reasons. First and foremost, Joe and I have known each other for a long time; he has been a very good friend; and I have known him as a serious person, honest to his conscience, who thinks deeply. The second and equally important reason is that I regard an understanding of sustainable development as the *overwhelming question* that we face today.

This phrase comes from T.S. Eliot's *Love Song of J. Alfred Prufrock*. in which he writes:

> "To lead you to an overwhelming question …
> Oh, do not ask, 'What is it?'
> Let us go and make our visit."

That visit would be a study of this very interesting book by Joe Hulse.

Dr. Hulse is a scientist, who has worked in industry, and with aid agencies, primarily in the promotion of research and development. His area of expertise is the life sciences, particularly relating to food, agriculture and agribusiness. He has frequently visited India and other developing countries. He was chairman of a committee (CASAFA) of the International Council of Scientific Unions (ICSU), when I was its president. This committee dealt with the application of science on an interdisciplinary basis in the area of food and agriculture, and particularly with reference to the developing countries.

Dr. Hulse has lived in a developed country, Canada; and there is a great deal of reference to Canada in the book, and to its distinguished

Prime Minister Lester Pearson, who was indeed one of the great leaders of the world. At the same time, Dr Hulse has great empathy with the developing countries and the poor. This partly arises from the many visits he has paid to India, and this empathy comes through in the book. There are also many references to the International Development Research Centre (IDRC), an organisation founded by Prime Minister Pearson and where Dr Hulse worked as a senior scientist from 1970 to 1988. As a former member of the Board of Governors, I am delighted that IDRC will co-publish this book.

The book covers a wide range of topics relating to sustainable development. I found it truly educative; as well as being easy to read and holding ones interest. It contains unusual nuggets of information in every chapter.

What sets the book apart is that there is, throughout its pages, in Joe's own words, a "philological and philosophical discussion of the etymological origins and changing concepts" of the different areas that he has dealt with. There are also accounts of the history of civilisation, tracing how we got where we are. The book represents the work of a true scholar and not of a technician.

Particularly over the past half-century, sustainable development has become a powerfully appealing idea. It has, of course, become most popular as jargon, following its usage by the World Commission on Environment and Development chaired by Dr. Gro.Harlem Brundtland, which published its well-known report in 1987 on *Our Common Future*. It is interesting that, in an ancient Indian religious writing over 2,000 years ago, it is stated:

"All in this manifested world, consisting of moving and non-moving, are covered by the Lord. Use its resources with restraint; do not grab the property of others, distant and yet to come."

This is the concept of inter-generational equity – respecting the interests of those "distant and yet to come" – that is central to the implementation of sustainable development. The concept is thus old; and a number of ancient references that lead to it can be cited from different parts of the world. At that time, with small populations and an abundance of resources in the ecological system, the problem was essentially ethical.

Today, there are many difficulties in our efforts to bring about sustainable development. These include: the inexorable growth of an already large population; the ethical dilemma where the rich do not want to share with the poor, and do not consider they have any responsibility to the latter; the driving forces of globalisation, the market economy and consumerism; and the visibly increasing pressure on the environmental life support systems. This brings out starkly the importance of ethical aspects in our functioning.

I am personally optimistic that we are moving forward. We are beginning to understand the importance of development in all its forms (political, social, economic, scientific, cultural, human); the links between development, poverty and peace; the power of science and technology to bring about a better life for the people; the scientific issues pertaining to the many areas where we could face major problems (climate, water, desertification, energy and the like); and the underpinning ethical issues. Most important, in all of this we are not dealing with wisdom handed down by revelation, but arrived at by tortuous processes of consultations involving stakeholders' interests, and large numbers of people.

To me the Brundtland Report has expressed this point very well: "Human survival and well-being could depend on success in elevating sustainable development to a global ethic." That is the need of the hour.

We have to modify consumption patterns. Mahatma Gandhi had made a perceptive point: "There is enough for everyone's need, but not enough for everyone's greed." It is possible to quote from all the great religions of the world (Hinduism, Islam, Taoism, Christianity, Confucianism, Buddhism, and the many more that are followed at various levels in society), which recognise restraint in consumption as a virtue. How continued economic growth, necessary for gainful employment, can be harmonised with this aspect of restraint on consumption will need to be worked out.

The book by Dr. Hulse provides us with a great deal of the background to deal with this overwhelming question. I would like to congratulate him on the major effort he has put in to clarify so many of the issues in this complex area in an eminently readable form.

The unwavering support of his wife Adeline has been important in making Joe as thoughtful, caring and productive as he has been; as a friend, I know this. My congratulations to Adie also.

Finally I am deeply touched by the sensitive dedication of this work to the memory of Professor Y. Nayudamma, who was also a dear friend of mine.

Professor M.G.K.Menon is currently Chairman of the Indian Institute of Technology (Delhi), Senior Advisor, Department of Space at the Indian Space Research Organisation, President of the Indian Statistical Institute in Kolkata, and Chairman of the Raman Research Institute Trust in Bangalore. Professor Menon has had a distinguished career as a scientist and policy-maker and has held a number of prominent positions including Minister for Science and Technology, Government of India; Secretary, Department of Science and Technology; Secretary, Department of Electronics; Member, Planning Commission; and Scientific Adviser to the Prime Minister of India. He is the recipient of prestigious awards such as Padma Bushan, Padma Vibhushan, and Abdus Salaam awards.

Preface

This book is about (i) 'sustainable development', a popular term often used to imply ethical, conservative and prudent utilisation of all resources and a compassionate disposition to all humanity and biodiversity; and (ii) individual and organisational failure to learn from past experience.

The words 'sustainable' and 'development' are discussed from an etymological and semasiological perspective since they are so often written about with no precise definition of what is meant or intended. Consequently the literature is replete with conflicting and confused concepts of what constitutes sustainable development. Comprehension of sustainable development demands first a systematic assessment and analysis of the many factors, influences and resources that impinge upon and interact in various and diverse developmental circumstances. There is no simple, single path to, or formula for, sustainable development. Each development must be considered in light of the prevailing social, economic, physical and, sometimes, political environment.

Few organisations, governmental and non-governmental appear to maintain a lively, up-to-date institutional or corporate memory: a critical, analytical record of past experience from which to define and formulate productive courses of action for the future, activities that avoid repeating errors of the past. As many wise philosophers have observed, failure to learn from history and past experience results in mistakes being repeated.

Equally, the text lays emphasis on the need for language that is precise, using words and sentences that convey an unambiguous meaning to all to whom they are addressed. In his 1946 essay on *Politics and the English Language*, George Orwell sensibly states that slovenly language leads to foolish thought and foolish action. The vast literature about sustainable development is replete with slovenly language: writers and development agencies speaking of sustainable development without ever defining what they mean.

The text reviews human development; the economic and environmental conditions, the resources and the biotechnologies: agriculture, food and health care, the diagnostics, prophylactics and therapeutics by which life on earth has over many centuries been sustained.

The literature related to international development – aid and assistance to poorer communities, the recorded history that describes the development of agricultural, food and pharmaceutical biotechnologies essential to human healthy survival, to human degradation of the earth's environment and ecosystems, is too vast to permit a definitive comprehensive review. The text attempts a record and commentary on what the author perceives as significant developments in:

a. international aid and assistance programmes;
b. the transition from artisanal empiricism to scientific understanding of biotechnologies essential to human health and welfare;
c. the patterns of change among industrial biotechnologies;
d. serious issues anent environmental degradation and climate change;
e. excessive exploitation of critical resources;
f. depletion of biodiversity, arable land and water, and
g. ethical and political issues that influence international trade, equitable distribution of wealth and equitable access to essential resources.

It begins with what is prescribed in the Atlantic Charter, the creation of the United Nations, its family of agencies and the international development banks. It reviews recommendations from various international commissions and conferences, from World Bank reports and UNDP Human Development Reports. It comments on governmental policies, human and industrial actions detrimental to the planet's environment and natural resources. It reviews the patterns by which biotechnologies essential to human survival and health have progressed over the past 6,000 years, and the consequences of uncontrolled urban growth on food and health security.

Responses to extensive disasters such as the Asian tsunami, to human suffering caused by atrocities in Rwanda and Darfur, suggest that civil society is frequently more ready and willing to act compassionately than are many governments and politicians, who are more intent on gaining and retaining power than in the survival and well-being of peoples who cannot vote for them.

The text illustrates the immense disparity between investments in armaments and commitments to programmes to alleviate poverty and disease. Several responsible observers are convinced that disease epidemics, atmospheric pollution and global climate change pose greater risks to human health and welfare than any conceivable forms of terrorism, that terrorism can be stimulated among people forcibly deprived of their land and property, among those who suffer extreme poverty and chronic unemployment.

Several observations are repeated in different chapters, primarily to emphasise their importance.

The essential thesis of the book is summarised in Titian's 'An Allegory of Prudence':

"Ex praeterito praesens prudenter agit
ni futura actione deturbet."
"From experience of the past
the present acts prudently
lest it despoil future action."

It is hoped the book will inform and be helpful to all who care about human suffering and degradation of the earth's environment and resources, in particular to men and women who are newcomers to international, governmental and non-governmental aid and assistance programmes. If the book encourages a more sensitive awareness of what has occurred in the past, of policies and priorities critical to ensure a peaceful equitable future for all humanity, for environmental stability and conservation of biodiversity, it will have served a useful purpose.

Joseph H. Hulse

March 2007

Acknowledgements

I gratefully acknowledge and thank the Board of Trustees of the National Gallery, London, for their kind permission to reproduce in reduced scale the Titian painting 'An Allegory of Prudence'.

I am deeply grateful to Professor M G K Menon for his extremely generous comments in the Foreword. His exceptional knowledge, unique and sensitive understanding of the subjects discussed, and his perceptive commentary add significantly to the book's content.

I offer my sincere thanks to Dr V Prakash and to Professor M S Swaminathan for their encouragement to expand into this book text a lecture presented to an international conference in 2004 at CFTRI, Mysore.

A particular expression of gratitude is extended to Dr B K Lonsane for his perceptive editing of the first draft text and for his constructive advice anent content and style of presentation.

I owe special appreciation to Clyde Sanger, a distinguished Canadian and international journalist, who composed the text of several case studies recorded in Chapter 10 and for his several suggestions related to content.

I am much indebted to Bill Carman, Publisher for the International Development Research Centre, for his patient, helpful advice and for his negotiations with the publisher.

My fondest appreciation goes to my wife Adeline who recommended inclusion of the Titian's 'An Allegory of Prudence'.

Finally, it must be placed on record that while many sources of information presented are referenced, for opinions expressed, several undoubtedly contentious, I accept full and sole responsibility.

Joseph H. Hulse

June 2007

1

Introduction

Sustainable development: scope and purpose of the text

The purpose of this book is to examine, in an historical context, sustainable development and the variant concepts that relate to:

a. international development: alleviation of poverty, deprivation, economic and social inequities, and what may influence future food and hygienic security for all humanity;

b. industrial biotechnologies as they have evolved and are applied to agriculture, food production, preservation and distribution, and pharmaceutical diagnostics and therapeutics;

c. concerns among civil society for their health and security in the light of exceptional innovations in agricultural and industrial biotechnologies, and the capacity of the planet's environments and resources to withstand increasing stresses imposed by human and industrial activities.

'Development' is a noun that may describe the state, pattern of change and progress over time of virtually every known phenomenon, human and industrial activity. It amounts to a contemporary act of faith that all 'development', for whatever purpose, must be 'sustainable'. 'Sustainable development' is a declared principle and purpose of international agencies, national governments and their ministries, industries – particularly those dependent on natural resources and biotechnologies – diverse organisations and civil societies concerned with human development, conservation of environments, biodiversity and critical resources.

The term 'Sustainable Development' gained currency during the 1970s. Since then, 'sustainable' has tended to imply that a process or activity will progress prudently and efficiently, with economic and benign use of resources; and that it will be unlikely to inflict insult or injury to the health and welfare of humans and other organisms, or damage to the ecologies and environments in which they exist. Sustainable development is susceptible to many and diverse interpretations, composed of complex components, some of which can be measured, determined or reliably observed and described, while others are derived from particular ethical principles, political or social ideologies, personal opinions and prejudices.

Difficulties of definition and concept

The text begins with a philological and philosophical discussion of the etymological origins and changing concepts of 'development' and 'sustainable', a discussion believed necessary in view of many vague and uncertain perceptions, confused and often conflicting objectives. Conceptual differences are evident in relation to 'international development': organised assistance to poorer nations and communities to raise their quality of life, employment opportunities and systems of governance; to agribusiness, agricultural and industrial biotechnologies; and to various bio-scientific research programmes.

Opinions differ on the probable consequences of interactions among environment, climate and demographic changes in relation to human, agricultural and bio-industrial activities; on issues relevant to international trade; options open to transnational corporations; and the potential influence of poverty and social inequities on civil unrest and international confrontations.

'Development', however it is defined, demands systematic analysis of the many complexities that impinge upon and influence economic, social and technological progress and change. It is axiomatic that fundamental scientific principles are universally valid and applicable, whereas technologies and political systems may not be readily transferable between nations and communities whose resources, constraints, traditions and social attitudes are distinctly different.

Corporate memory

A review of international development activities since World War II indicates that few development agencies maintain a comprehensive and reliable corporate memory, and that many administrators appear relatively unaware of what has gone before. The Book of Ecclesiastes cogently observes: "There is no remembrance of former things". Cicero wrote: "Not to know what has been transacted in former times is to continue always as a child". George Santayana, the Spanish-born philosopher, in his *Life of Reason* comments that those who fail to learn from history are prone to repeat the mistakes of the past. An anonymous writer wisely notes: "Historical memory is the only dependable guide to rational navigation into the unknown waters of an uncertain future". As is illustrated in many of the publications cited in this book, the future of development, however defined or conceived, is uncertain. All persons and organisations that seek to devise, design or embark on 'development' programmes would therefore be wise to learn from past experience.

In the later text, international development is reviewed historically starting with the Atlantic Charter, the creation of international development banks, the United Nations and its agencies. There follows a review of the many conferences, commissions and publications that have examined past progress and have prescribed priorities and programmes of action to alleviate poverty and malnutrition while protecting and conserving resources essential to the survival and satisfactory existence of future generations. Emphasis is given to countries that have progressed from relative poverty to comparative prosperity; explaining how other countries have fallen behind, and describing the immense economic and social inequities in access to basic necessities and opportunities within and among the world's nations.

Indicators of human and communal development have evolved from a simple economic index (Gross National Product *per caput*) to other more complex indices that include health, education, social conditions, ecological and species protection, and systems of governance. The book will trace agricultural and urban development from the ancient river-valley civilisations to modern mechanisation and genetic modifications, how agribusiness and pharmaceutical industries have progressed from artisanal processes, which emerged from perceptive empiricism, to biotechnologies derived from systematic scientific research and development. It will define and discuss biotechnologies (plural) more comprehensively than the prevalent contemporary disposition to speak of biotechnology (singular) as solely confined to genomics and the genetic modifications of botanical and zoological species.

Ancient concepts of sustainable development

The concept that development, for whatever purpose, must be long-lasting and not ephemeral, and that present demands and consumption must not deprive future generations, goes back many centuries. In the third century CE*, Tertullian wrote: "Our numbers are burdensome to the world, which can hardly supply us from its natural elements as our wants grow more and more demanding".

---

* Throughout the text, where a date is cited, 'BCE' (Before the Common Era) or 'CE' (Common Era) are used instead of BC and AD respectively.

If the biblical Joseph ever existed, his decision to store grain harvested in years of abundance to provide for later years of scarcity was probably inspired less by imaginative dreams than by familiarity with records maintained by a succession of Egyptian dynasties who meticulously recorded the time each year when the Nile reached its maximum flood, and the years when it failed to flood because of drought in the Ethiopian highlands from which the Nile waters are fed. During lean years, with too little water for crop irrigation and without access to stored grain, there would be famine in the land [de Burgh 1947]. From their scrupulous observations, the Egyptians noted there were roughly 365 days between each maximum flood, from which observations they devised a calendar several millennia before the Julian calendar.

Archaeologists and Egyptologists have discovered and deciphered documents that include the teachings of Kagemna (*ca* 2900 BCE) and Amen-em Apt (*ca* 1500 BCE) and the rules of governance enacted by Amenhotep [Ikhn-aten] (*ca* 1400 BCE), all of which prescribe ethical patterns of government and individual responsibility to take care of the poor and the sick. In the Epilogue and several of his 282 codified laws, Hammurabi, King of Babylon (*ca* 2200 BCE) decreed that there be justice and fair treatment for all his citizens, penalties for those who oppress or cheat their neighbours and long-term provision for the needs of poor widows and orphans [Harper 1999]. Deploring the greed of Roman merchants, Cicero insisted that the State take care of and provide for the safety and security of all Roman citizens, both present and future. Cicero also believed that awareness of past history and experience would inspire moral standards of governance to ensure justice for all citizens, those living and others yet to be born [Cowell 1948].

In the seventh century CE, the Prophet Mohammed inspired the rise of Islam (Surrender to God) because of his disgust with wealthy avaricious Arabian merchants. An essential pillar of Islam is "Zakat", an obligation imposed on all Muslims to give a significant proportion of their wealth to the poor [Armstrong, K 1994].

Development since World War II

The conscientious duty of the privileged and powerful to be compassionate towards the poor and disadvantaged is a doctrinal edict of most major religions. Individual and institutional acts of benevolence have a long and heart-warming history. Following World War I, the League of Nations was created with noble intentions. But it is only since World War II that development, specifically intended to relieve poverty, chronic disease, economic and social inequities, has become truly international. A review of aid agencies' programmes and priorities illustrates how patterns and purposes of developmental assistance have changed, how economic and industrial progress in much of Asia contrasts with persistent poverty and deteriorating hygienic, economic and social conditions among nations of sub-Saharan Africa. Particular attention is given to agriculture, agribusiness and biotechnologies that furnish food on which healthy survival depends, and drugs designed to diagnose, prevent and cure acute and chronic maladies. The text examines contrasting opinions and predictions for the planet's environment, resources, atmosphere, climate, arable land, fresh, inland and coastal waters, forests and biodiversity.

The book addresses the miseries inflicted by egocentric ideologies that are claimed to be divinely dictated and imposed on others by force. It illustrates the advantages of south-south cooperation between and among nations at different stages of economic and technological development, as opposed to the tied aid policies and practices of donors' bilateral programmes.

Food systems analyses

Uncontrolled growth of urban communities adversely affects resources and facilities critical to future food security. There is an urgent need to determine future demand, supply and distribution of food to expanding urban communities; to design comprehensive systems in which rural crop and livestock production are holistically integrated with reliable post-harvest preservation and economic distribution from rural producers to urban processors and markets. There is need and opportunities for mathematical simulation models to determine the most efficient channels and times for rural to urban and intra-urban transportation of perishable foods. Opportunities are apparent for employment in rural agribusiness, in primary processing of perishable food sources to satisfy changing demands of urban consumers. Professionally competent systems analysts are urgently needed to cooperate with scientists engaged in research and development for agricultural and industrial biotechnologies. The author chairs an international working group that studies and recommends on holistic integrated food systems. The group's aim is to help ensure food security for nations and communities at various stages of development and need, nations with access to a wide diversity of resources.

Past experience in farming and food systems analysis illustrates how development agencies and governments must first analyse and understand what of relevance exists before prescribing potential improvements. Food systems analysis first takes account of the opportunities and resources needed for productive development, the constraints and risks contingent on adaptation and adoption of unfamiliar innovations and novel technologies.

Industrial biotechnologies

Throughout the world the need and demand for improved health care are growing and diversifying. In consequence, health care industries are changing in structure and activities. The cost and time to discover and develop new drugs are becoming prohibitive for even the largest corporations. The pharmaceutical industry is reforming into two complementary sectors: the large drug companies supported by hundreds of 'specialist biotechnology' companies that provide specific services under licence or through other cooperative arrangements with larger corporations or other health care providers. The demand for health care services may soon be greater than can be sustained, particularly where governments seek to provide comprehensive health care free of charge for all people.

Ethical, ideological and political issues

Ethical and political issues discussed here relate to

a.   bioscientific innovations, such as transgenic modifications, human embryonic stem cells and functional genomics;

b.   systems of international and national governance;

c.   integrity in corporate management among transnational companies;

d.   international trade in products of agriculture and biotechnologies;

e.   inequities that result from massive annual subsidies to agricultural and other commodities by European and North American governments;

f.   failure among affluent nations to legislate for fair and equitable trade as prescribed in the Atlantic Charter.

It is suggested that transnational corporations at all times behave as responsible corporate citizens, not as commercial colonisers.

It is strongly recommended that donors' tied-aid policies give way to assistance that enables recipients to develop and adopt technologies and systems that are appropriate to their needs and resources. Much development investment has been wasted by sending 'experts' with little prior understanding of the social conditions, cultures and traditions of those they are sent to assist; by transferring technologies ill-suited to the needs and resources of aid recipients. Nations such as India, that have made impressive agricultural, technological and industrial progress, can provide more productive and appropriate assistance to sub-Saharan and other poor nations than tied-aid programmes imposed by affluent donor governments.

The text addresses the concerns of consumers and civil society about safety and possibly adverse consequences of modern biotechnologies used to produce foods and drugs, with particular reference to modern genetic modifications and human embryonic stem cells. The rising demand for, and costs of, providing health care for all humanity will be related to the future priorities and resources of governments, international agencies and the private sector.

General comment

It is recognised that the scope essayed is ambitious and wide-ranging, and that much of relevance is perforce omitted. It is emphasised that the opinions offered, some adversely critical, are solely those of the author. Having devoted more than 40 years to international development activities, the author fully understand that the difficulties are diverse and complex, and that there are no simple solutions applicable to all communities and circumstances.

It is hoped, however, that the observations offered will encourage all who become engaged in 'development' to define, first and in precise terms, their objectives and criteria of assessment, and to analyse critically and understand comprehensively what already exists, and what has taken place in the past – before seeking to bring about beneficial change.

2

Definitions and Concepts

Defining and determining development

The eminent physicist Lord Kelvin contended that "if you can define and measure precisely that of which you speak, you know something of your subject; but if you cannot, your knowledge must be considered meagre and unsatisfactory." Antoine-Laurent Lavoisier, the great eighteenth century French chemist, insisted that scientists be as accurate and precise in their speech and writing as in their analytical measurements. In Lewis Carroll's *Through the Looking-Glass*, Humpty Dumpty, after speaking semantic nonsense, defended it by saying: "When I use a word it means exactly what I want it to mean, neither more nor less". The popular non-scientific literature concerning 'sustainable development' at times seems more in tune with Humpty Dumpty than with Kelvin and Lavoisier.

The Pearson Report [Pearson 1969] stated that 'development' presented his Commission with exceptional semantic difficulties. The Brandt Commission [Brandt 1980] reported it would not try to define 'development'. The Brundtland Commission [Brundtland 1987a,b] made no attempt to define 'development' but simply stated that "development is what we all do in attempting to improve our lives".

The word 'development' came into the English language around 1750, since when it has been applied to suggest patterns of change and progress in virtually every form of human, corporate and institutional activity. From the late 1940s 'international development' gained currency with the creation of the International Bank for Reconstruction and Development (the World Bank), the United Nations Development Programme, and, among the industrial nations, establishment of government-financed 'development agencies' devoted to offering technical, social and economic assistance to poorer nations in Africa, Asia, Latin America and the Middle East. As Pearson, Brandt and Brundtland indicate, 'development' in this context defies precise definition and determination, as is evident from discussions about development.

The Oxford and Webster dictionaries define 'development' as "a gradual unfolding", suggesting that the word is derived from the Latin *"de"* and *"volutus"* meaning "to unveil". Salome's dance of the seven veils is illustrative of a gradual unfolding. The concept of gradual unfolding deserves serious consideration by agencies that seek quick solutions to highly complex difficulties. During the Bretton Woods Conference in 1944, Harry Dexter White, a senior official in the US Treasury, stated emphatically that to be effective and long-lasting, all development requires gradual, cautious planning and patient, persistent commitment. In other words development is most reliable when it unfolds gradually. A similar sentiment is expressed by Pearson, Brandt and Brundtland who collectively assure us there are no rapid remedies for chronic poverty, malnutrition and inequities between rich and poor. Brandt states specifically: "Quick solutions are an illusion"; Brundtland insists there are "no quick-fix solutions".

What is sustainable development?

The broad concept of sustainable development is often attributed to the Brundtland Report which states: "… we must ensure development meets the needs of the present without compromising the ability of future generations to meet their needs" and adds: "Sustainable development is not a fixed state of harmony, but rather a process of change in which the exploitation of resources, the direction of investments, the orientation of technological development and institutional change are made consistent with both future and present needs". The Report goes on to say: "... the development process is not easy or straightforward ... and in the final analysis sustainable development must rest on political will". Sadly, political will and commitment are often of short duration, without the persistence essential to ensure that development gradually unfolds until precisely defined objectives have been realised. As political and administrative leaders change, so do donors' policies and priorities.

Writing in *Harper's Magazine* in January 2006, Paraj Khanna dismisses 'political will' as an oxymoron. Political policies change as governments and their heads of ministries change. For the most part, political will and priorities reflect the interest of politicians and those who support them.

The English verb 'to sustain' and its archaic predecessor 'to sustentate' are derived from Latin verbs *'sustentare'* and *'sustinere'* which are defined as 'to hold up', 'to support' or 'to sustain'. It is interesting, in the context of sustainable development, that a related Latin noun *'sustentatio'* is defined as 'patience'. 'To sustain' and its Latin antecedents are active, transitive verbs, they require both a subject and an object: something or someone must act to sustain whatever is to be sustained.

This author has yet to find the noun 'sustainability' in any standard dictionary and agrees with Charles Kidd of the American Association for the Advancement of Science who wrote: "The roots of the term 'sustainability' are so deeply embedded in fundamentally different concepts ... that a search for a precise definition seems futile. The existence of the word is tolerable only if each analyst describes clearly what he or she means by 'sustainability'." Very few do so. [Kidd 1992]. In a scholarly and thoughtful publication Kidd considers the literature related to sustainable development to be so voluminous that any attempted comprehensive review and analysis would be impracticable and the outcome not worth the effort. Kidd reviews successive 'movements' that have prescribed and proposed means to realise sustainable development.

Sustainable: static or dynamic?

Some concepts of sustainable development suggest a condition comparable to static hibernation, or a state of perpetual motion with no increased demand on resources. During the fifty years after Rudolf Clausius, the mathematical physicist, formulated in 1850 his second law of thermodynamics, no less than 500 applications claiming inventions that promised perpetual motion were received by the British Patent Office. 'Sustainable' can be literally interpreted as an unchanging state or condition, actions taken to prevent a system or entity from collapsing. Termites have sustained their particular patterns of social organisation and habitation for more than one million years. They used evaporative cooling to control the internal temperature of their habitats thousands of years before James Joule and Lord Kelvin published their classical studies on expansion of gases and explained the physical principles of evaporative cooling.

But in most instances development proceeds under conditions that are changing. Whether or not the rate and pattern of developmental change are sustainable depends upon how effectively the development process is controlled, how well it follows a predetermined path. Planned sustainable development requires that the intended objectives and the criteria by which progress is to be assessed are precisely stated. All too often development is claimed to be sustainable when neither objectives nor criteria of assessment are declared, where there is little evidence that progress is systematically determined.

'Sustainable development' has come to suggest an assurance of prudent, conservative practices, of ethical integrity and respectability. But many claims of sustainable development do not bear close scrutiny. Furthermore, what may truly be sustainable in one location or community is rarely applicable and amenable to all locations and communities. Systems acceptable and affordable among the affluent are often beyond the resources of poorer people. It is not unusual for people whose lives are not at risk from chronic poverty to recommend to poorer nations remedies for what are essentially the environmental sins of Europeans and North Americans. A thoughtful publication by the Canadian Agricultural Economics and Farm Management Society states that the published literature on sustainable development indicates inadequate investment in human capital, legal infrastructures, technology and physical capital, and the substantial role such investments play in the economic evolution of human societies [CAEFMS 1990].

Ubiquity of 'development'

As stated in the Introduction, 'development' is a noun that may describe the state, pattern of growth or change of progress or retrogress of virtually every known phenomenon and activity. 'Development' can relate to the cosmic universe, the earth's environments and ecologies, global, regional or national economic, agricultural and industrial progress and the many and diverse components thereof. A Canadian foundation, Sustainable Development Technology Canada (SDTC) describes its mission: "The Foundation will act as the primary catalyst in building a sustainable development technology infrastructure in Canada". It defines sustainable development technology as "Products that integrate economic viability, environmental stewardship and social equity, and meet the needs of the present without compromising the ability of future generations to meet their own needs", a concept borrowed from the Brundtland proposition. Without in any sense seeking to derogate the worthy intentions of SDTC, from present industrial practices, particularly those of the petroleum producers and high energy consumers, there is little consensus among Canadians and other North Americans as to what constitute sustainable development technologies as defined by SDTC. Nor among politicians is there much concern for the needs of future generations, whether they will aspire to a standard of living and resource utilisation comparable to those of contemporary North

Americans, to higher levels of consumption, or to lifestyles less demanding on commercial energy and other resources.

Primitive and early human development

During humanity's earliest recorded existence, development was synonymous with survival. Each family or primitive community survived and developed by finding access to resources sufficient to feed, house and clothe itself. Simple tools and weapons provided means by which a family could sustain itself and defend its persons and possessions from enemies. Some communities survived and developed, others became extinct. Life was hazardous, brutish and of short duration. As hunters, gatherers and primitive communities coalesced, first into city-states, later into nations, development became the responsibility of those who governed. Ancient Egyptian, Babylonian and Harappan civilisations progressed from subsistence to commercial agriculture, enabling fewer farmers to provide sufficient food for others to engage in non-farming activities; and to carry over surpluses from regions and seasons of abundance to those of scarcity.

Nations that increased in size and power conquered and dominated weaker nations and developed empires, conquered peoples being forced into servile slavery for the benefit of the conquerors. Among nations that seek to dominate others, development objectives tend to be defined in terms of economic and military power.

Early military power was later enforced or superseded by the domination of religious organisations. The Holy Roman Empire, administratively modelled on the Roman army, was dominant across Europe for more than 1,000 years. Religion allied with military muscle enabled powerful European nations to colonise large areas of Africa and the Americas. Islam followed and expanded, first by domination of the Semitic nations, later spreading its influence to other continents. Starting after World War I, political ideologies, particularly those derived from Marx and Lenin, dominated huge regions of Eastern Europe and gradually influenced politicians in other lands. National socialism enforced by Germany and Italy during the 1930s, though of shorter duration, affected the lives of many millions.

Though colonisers of all ambitions and persuasions – military, religious, or ideological – claim to be first and foremost concerned to improve the wellbeing of the minds, souls and bodies of their subjected adherents, in practice their criteria of development are dictated by the pursuit of power. How successfully they expand and maintain their controlling influence determines their self-perceived state of development. At the peak of its magnitude and prosperity, it was claimed that the sun never set on the British Empire.

Development out of poverty

For as long as there have been historical records, rulers and philosophers have proclaimed the duty of the rich and powerful to refrain from oppressing the weak and to be charitably compassionate to the poor. However, only since World War II, following the gradual emancipation of former colonies, has development with the objectives of alleviating abject poverty, enhancing the livelihood opportunities and ensuring food and health security for all humanity become internationalised. International development: assistance from the more affluent to poorer weaker nations is now administered among various international, national and non-governmental organisations, among United Nations' agencies, cooperatively and collectively, by governments and NGOs on an individual or bilateral basis. At first, results of international development programmes were judged by relatively simple parameters such as *per capita* growth in Gross National Income. Subsequently, international development objectives have become more complex and by no means uniformly understood or sustained by all development agencies.

Thus 'development' can pursue various objectives, and be judged by diverse criteria. As applied to so great a diversity of phenomena, activities and processes, 'development' eludes any simple definition or description. Each intended development must be defined, determined and assessed by specific criteria, some that can be quantified or reliably observed, others dependent on subjective philosophies, persuasions, opinions, ideologies and prejudices. When the adjective 'sustainable' is applied to qualify the noun 'development', definitions, criteria and methods of assessment dilate in complexity.

Complexity of international development

International development agencies came into being soon after 1945 to assist poor nations, many transformed from colonies into independent sovereign states. Declared development objectives were for the newly independent states to be governed efficiently and economically, for their inhabitants to enjoy freedom from poverty, hunger and insecurity.

Human development

'Human development' programmes are generally designed to raise the quality of human life. In their book *Only One Earth* [1972] Barbara Ward and René Dubos describe human development, in relation to the conditions and environments in which people live, as "a very subjective and ill-defined concept". They discuss a conceptual philosophical difficulty that confronts those who seek to rationalise the interactions of human activities with other living organisms. They raise the question: Are humans simply higher apes as their genes suggest, and therefore of no greater importance or significance than other living organisms, or should *Homo sapiens* be considered a unique and discrete genus and species that deserves a distinct and special place among all creatures great and small? The Book of Genesis states that God created man in his own image, both male and female, and said to them: "Be fruitful and multiply, replenish the earth and subdue it; have dominion over the fish of the sea, over the fowl of the air and over every living thing that moves on the earth."

The proposition that *Homo sapiens* enjoys divinely ordained privileges and domination over all other creatures is an issue of conflict between those of certain religious beliefs and others who defend the rights of animals and urge benign conservation of biodiversity. The UN Conference on Environment and Development (Rio, 1992) proclaimed in Principle 1: "Human beings are at the centre of concerns for sustainable development. They are entitled to a healthy and productive life in harmony with nature." Principle 7 proclaims: "To achieve sustainable development and a higher quality of life for all people, States should reduce and eliminate unsustainable patterns of production and promote appropriate demographic policies." While few would disagree with the underlying principles, these noble proclamations are open to widely different practical interpretations.

Indices by which to assess human development continually evolve and vary among analysts and interested agencies. Since a primary objective is to alleviate poverty, estimates of national and individual incomes and annual rates of growth are commonly factored into human development indices. Estimates of individual incomes are at best approximations, many being calculated as the quotient of a nation's total income divided by the numerical size of its population. Population and national income statistics generally compiled by governments, across nations are by no means uniform or consistently reliable. The ability to collect and analyse accurate economic and demographic data varies among nations, the processes being difficult and expensive. Nor are governments always willing to share with development agencies economic information that reflects unfavourably on their management efficiency.

Indices of economic development

'Developing countries' is an accepted term to describe the poorest nations. The World Bank sub-classifies developing countries into 'Low Income Countries' (LICs), 'Lower and Upper Middle Income Countries' (LMICs and UMICs). Indices of national annual income include 'Gross Domestic Product' (GDP) expressed in US dollars at current rates of exchange, defined as the sum of the value of all output produced by economic activity within a country during a particular financial year. Estimates of GDP may be based on market prices or factor costs, the economic activities considered appropriate for inclusion being variable among government economists. Consequently, GDP quotations must be considered as estimated indicators rather than precise statistics. 'Gross National Product' (GNP), also alternatively based on market prices or factor costs, is defined as GDP plus net income earned abroad. Reservations similar to those cited anent GDP apply to estimates of GNP. Command GDP, a recently added concept, includes what economists describe as "terms of trade", such as the relative values of imports and exports. Between 2000 and 2005, conventional GDP for Canada rose by 15 per cent, whereas Command GDP increased by 18.2 per cent.

Since GDP and GNP are customarily quoted in US dollars, the values are influenced by prevailing rates of exchange between national currencies and the US dollar. The local market prices of foods and other goods essential to human health and welfare differ significantly among nations and over time. GDP/Cap and GNP/Cap expressed in US dollars do not reflect comparative costs and standards of living, or the relative purchasing power of families and communities in different locations. To adjust for cost of living disparities among nations, economists have devised an indicator known as 'Purchasing Power Parity' [PPP] in which official exchange rates between currencies are adjusted to equalise prices of a standard 'basket' of goods and services. Average GDP/Cap expressed as PPP is calculated on a scale of 1 to 100 where the United States, considered to enjoy the highest standard of living, is rated at 100. Other nations' PPPs are calculated using the number assigned on the PPP scale. Calculations for the year 1999 illustrate the World

Table 1
GDP/Cap and PPP/Cap ($USD) for 1999

	(a)	(b)		(a)	(b)
OECD	22,000	28,500	Ethiopia	100	640
Canada	23,000	28,000	Burundi	110	600
U.K.	22,000	24,000	Tanzania	270	535
Asia	2,700	4,300	Nigeria	280	800
India	450	2,230			
P.R. of China	800	3,600			

Bank's estimated differences between average GDP/Cap at (a) official rates of exchange, (b) at PPP adjusted GDP/Cap for several nations and groups of nations.

The actual compositions of standard baskets vary significantly among nations and communities; the dominant food preferences among inhabitants of Mumbai and Montreal are widely different. Consequently, PPP comparisons are best estimates rather than exact determinations. They are, nonetheless, more useful comparative indicators than GDP expressed at official rates of exchange.

In a further attempt to standardise the 'basket', *The Economist* journal has devised Big Mac Indices: the price charged for a Big Mac hamburger sold from McDonalds' fast food outlets in different nations. In *The Economist*'s opinion the Big Mac is one of the few foods of relatively standard composition sold in outlets around the world. In January 2004 *The Economist* determined that the price of a Big Mac in Beijing was $1.23 USD compared with $2.80 in US cities, from which it concludes that the yuan was undervalued by 56 per cent against the US dollar. Other Big Mac Index comparisons indicate to *The Economist* that in January 2004 the Euro was overvalued by 24 per cent, and the Japanese yen undervalued by 12 per cent.

Continuous growth in income, saleable and negotiable assets has long been regarded as a determining indicator of human, national and industrial development. For many generations poverty was determined in terms of estimated income. People rose out of poverty as their incomes increased to exceed some arbitrarily decided sum. Progress and changes in monetary status remain as determinants in assessment of development. But in recent years various social indicators have been added to and integrated with monetary parameters to broaden concepts and estimates of human development.

Social indicators of development

In 1987 the World Bank compiled a set of social indicators related to human development. The relevant publication opens with the following statement: "The ultimate goal of development is to improve the human condition, especially for the least privileged members of society. To this end, the World Bank, working with national governments, seeks to raise general living standards and reduce mass poverty ... As a means of assessing progress towards this goal the Bank has compiled a variety of social indicators incorporating data from the countries themselves in accordance with national definitions, or from specialised international agencies which deal with education, health and nutrition, and other social sectors... Economic development causes the supply of, and demand for, material goods to increase. But social dimensions of life also change. Some see this as a *result* of economic change: as economic wealth increases, population patterns change and institutions expand to deliver a more diversified range of social services. Others see these social changes as *causes* of economic change: for example, educational improvements help raise economic productivity. From either perspective, indicators of social change can be used to complement economic gauges as indicators of development and in formulating development policies ... Reaching agreement on what indicators signal higher social development and how to measure them is not simple ... definitions and methodologies vary among countries, readers are urged to use caution in interpreting the data." [World Bank 1987]

The report begins with summarised tabulations comparing selected social indicators for the years 1965 and 1985, then devotes two pages of available indicators for each nation classed as LIC or MIC in alphabetical order, from Afghanistan to Zimbabwe. The selected indicators include GNP/Cap, per cent of labour force employed in agriculture, infant mortality (per 1000 live births), average kCals/cap, per cent of population enrolled in primary school education. In Asia, Africa and Latin America, between 1965 and 1985 infant mortality fell significantly, primary school enrolment increased, and calories/cap rose in all regions except Africa where the average dropped to barely 2000 kCals/cap/day.

In the *World Development Report* for 2003, devoted to Sustainable Development, the World Bank's key indicators of development include Gross National Income (GNI), GNI in Purchasing Power Parity, rate of growth in GDP/cap, life expectancy at birth, mortality rate per 1000 under 5 years of age, per cent of illiterate people over 15 years of age, and carbon dioxide emissions (millions of tonnes/year). The gradual expansion of development to include economic, social and environmental indicators, while eminently justifiable, has made assessment methodologies more difficult and complex.

UNDP Human Development indices

In 1990 Dr Mahbub ul Haq, former Minister of Economic Development in the government of Pakistan, became the principal author of the first in a series of Human Development Reports published by the United Nations Development Programme (UNDP). Comparative national ratings were based on a novel indicator: the Human Development Index (HDI).

'Development' was earlier conceived in terms of national GDP growth, it being assumed that the benefits of increasing economic growth would trickle down to poorer members of society. Mahbub ul Haq insisted that, while economic growth is essential, high growth rates do not automatically translate into higher levels of human development. Alleviation of poverty, caused by too little access to income, assets, credit, social services and employment opportunities, is important but not the sole factor in human development. HDI takes account of life expectancy, adult literacy and standard of living, and is calculated from several sub-indices: a. Human Poverty Index (HPI), b. Gender-related Development Index (GDI), and c. Gender Empowerment Measure (GEM).

Included in the HDI computation are estimates of poverty (per cent of population that survive on less than \$1.00/day), proportion that are malnourished, per cent enrolled in primary education, gender equality or inequality, child mortality, access to clean water and hygienic sanitation. Considerations of gender equality include comparative estimates between males and females of life expectancy, adult literacy, earned income, control of economic assets and resources, and political participation.

A perfect HDI equals 1.000. Table 2 presents estimated HDIs for several nations for 1980 and 2001. Given the complexity of the calculation, there is probably no significant difference among highest rated affluent nations.

Table 2
HDIs for selected nations, 1980 and 2001

	1980	2001
Canada	0.881	0.937
USA	0.883	0.937
UK	0.847	0.930
India	0.443	0.590
P.R. China	0.554	0.721
South Africa	0.676	0.684
Zimbabwe	0.570	0.496

With the exception of sub-Saharan Africa, and at variable rates, over the past decade HDIs have risen across all developing regions. The following tables display the HDI rankings of the highest, medium and lowest nations and regions for the year 2003, as stated in *HDR 2005*. The number indicates their order or ranking: e.g., Norway topped the list with the highest HDI, while Niger was the lowest.

No	Highest	HDI	No	Medium	HDI
1	Norway	0.963	62	Russian Federation	0.785
2	Iceland	0.956	85	P. R. of China	0.795
3	Australia	0.955	120	South Africa	0.658
4	Canada	0.949	127	India	0.602
10	USA	0.944	135	Pakistan	0.527
15	UK	0.939	139	Bangladesh	0.520

No	Lowest	HDI	Regions	HDI
161	Eritrea	0.444	Least Developed	0.518
170	Ethiopia	0.367	East Asia-Pacific	0.768
176	Sierra Leone	0.298	Latin America –Caribbean	0.797
177	Niger	0.281	South Asia – Sub-Saharan Africa	0.515

Development: Controlled or unconstrained?

In light of the rapid growth and diversification in industrial and technological activity several thoughtful people question whether human beings possess the knowledge, the power and the wisdom to determine their destiny, the destiny and needs of future generations. Is the human race willing and able to establish a stable relation among themselves, the environment and all other organisms? Does the most affluent nation's pursuit of greater material wealth and military power forebode extinction of much that now exists and severe deprivation for many millions of less affluent and less powerful people now and in the future? Is it inevitable that extremists of various religious ideologies condemn the rest of humanity to endless conflict and suffering?

In 1798 the Rev. Thomas Robert Malthus published *An Essay on the Principle of Population as it Affects the Future Improvement of Society*. It provoked a prolonged and heated controversy that is by no means dead and buried. Malthus presented the opinion that human populations, if unchecked by disease, sexual abstinence or military conflicts, would inevitably increase more rapidly than the means of subsistence. A few years earlier Benjamin Franklin voiced similar apprehensions: "Whenever we attempt to amend the scheme of Providence, and to interfere with the government of the world, we have need to be very circumspect lest we do more harm than good." About fifty years ago, when populations of Asia, Africa and Latin America were rapidly increasing there were many who expected Malthus's pessimistic prophecies to be fulfilled.

Resource conservation

During the early period after World War II, evident among the conservation movement was Malthusian concern that the Planet's resources were inadequate to sustain the rapidly expanding economies among industrial nations, or sufficient to feed fast increasing populations of Africa, Asia and Latin America. In addition to concerns about resources needed to sustain and satisfy diversifying demands of increasing populations, evident was a rising insistence that more attention be given to environmental conservation. During the 1960s and 1970s, perhaps inspired by Rachel Carson [Carson 1962], 'the environmental movement' and 'the limits to growth movement' gained momentum. The environmental movement in the 1970s expressed frustration with big business, big government, and the pervasive pursuit of material affluence irrespective of the consequences for the global environment. Ecologists contended that failure to control human, industrial and other development activities would result in ecological disaster. One publication, based on proceedings of a Conservation Foundation Conference, stated: "... international development to date has been destructive." [Farrar and Milton 1972]. The publication warns of "the stressing of natural systems beyond their capacity for regeneration", a forerunner of more recent concerns about air and water pollution, land degradation and depredation of biodiversity.

The pursuit of excessive wealth and material possessions by powerful persons and political regimes has troubled moral philosophers for many centuries. Hammurabi, King of Babylon some 4,000 years ago, in the Epilogue to his *Code of 282 Laws* emphasises how his rules of governance prevented the strong from oppressing the weak. Several of Hammurabi's Laws prescribe penalties for farmers whose careless practices cause damage to their neighbours' lands and resources [Harper 1999].

The 'No growth-Slow growth' movement generated several challenging publications. One set of authors connected with the Club of Rome was persuaded that by the end of the twentieth century the world's economies were doomed to collapse because of rapid population growth, depletion of critical resources, degradation and pollution of the environment, and excessive resource demands of the wealthy and powerful. These conclusions were derived from several alternative computer simulation models, each of which forecast global disasters unless economic growth and individual and corporate resource demands and exploitations were severely restricted [Meadows *et al* 1972, Meadows 1977]. Advocates of these movements insisted that industrialised nations must enforce energy conservation and replace fossil fuels with renewable alternatives.

'The eco-development movement' advocated harmonisation of economic objectives with ecologically sound management. Robert Riddell, a professor at Auckland University, New Zealand, and a planner and ecologist, predicted in his 1981 publication *Eco-development, Economics, Ecology and Development*: "Now that oil wells are starting to run dry, northern nations will rattle their nuclear sabres in an effort to capture the precious remainder" – a perceptive prophecy of the US-UK invasion of Iraq. It is not surprising that affluent political jurisdictions oppose any suggestion that their acquisition and exploitation of Planet Earth's resources be restricted. They contend that research and improved technologies will provide adequately for the future, or, if necessary, the needed resources can be acquired from defenceless communities by military muscle.

Protection of the environment

Since it came into being in 1948, the International Union for the Conservation of Nature (IUCN) has campaigned vigorously for nations to be more protective and conservative of their own and global environments. IUCN was among the first to relate 'sustainable' to human welfare and the environment. IUCN's *1972 Yearbook* encourages conservation of environmental resources and pleads for political decisions and actions to assist all to realise the highest sustainable quality of human life. IUCN encouraged other agencies to pursue the principle that sustainable growth must take account of the limited availability of the Planet's natural resources. All development plans should assess potentially adverse environmental consequences, and be sensitive to the needs of future generations. Intergenerational equity received its most emphatic boost from the Brundtland Report [Brundtland 1987a,b].

Before more comprehensive concepts arose, environmental degradation was largely associated with atmospheric pollution caused by the combustion of fossil fuels that energised industries and motor vehicles among affluent industrialised nations. Many poorer nations, however, regarded industrial growth as essential to economic development. There is the often repeated story, probably apochryphal, of an African head of state who, when asked what he hoped to see as a stimulant to his country's economic future, replied: "Lots of smoke pouring out of lots of factory chimneys."

The 1972 UN Environment Conference in Stockholm described abject poverty as a contributor to and consequence of environmental degradation, specifically mentioning inadequate and unhygienic sanitation, contaminated water, squalid slum housing, natural resource depletion such as slash-and-burn agriculture and cutting of trees for fuel, overgrazing of natural pastures and loss of arable land. The United Nations Environment Programme, an outcome of the Stockholm Conference, the World Watch Institute, the World Resources Institute, an increasing number of environmentally sensitive NGOs, and the 1992 UN Conference on Environment and Development at Rio collectively excited public and governmental consciousness and concern for environmental degradation resulting from diverse patterns of development. Opinions remain polarised between a. biologists, ecologists and conservationists who foresee deprivations for future generations if present consumption patterns are unconstrained, and b. those more affluent who remain content to persist in their extravagant exploitation of natural resources, in the stated belief that science and technology will satisfy all future needs; and who, despite evidence to the contrary, insist that pollution of the atmosphere, surface and ground waters, and of coastal ecologies pose no serious obstacle to unrestricted economic and industrial expansion.

Scientific and ethical issues

In a thoughtful paper, scientists from the Italian National Institute of Nutrition and the International Foundation for Ecological Economics discuss scientific and ethical assessments of sustainable development [Giampetro and Bukkens 1992]. The authors emphasise the complex multidisciplinary nature of sustainable human development and the need adequately to protect the biosphere. They describe energetic models by which to study interactions between human societies and their environments, models that cover a range of time-scales, from days to millennia, over which to assess human exploitation of 'biophysical capital', a term proposed to describe the capacity within ecosystems to convert solar energy for the generation of biophysical processes that stabilise the biosphere's structure and function. The authors estimate that the standard of living in the United States requires a fossil fuel energy expenditure of 200,000 kcals/day/person, an extravagant level of consumption unavailable to, and unsustainable for, most of the rest of the world, particularly poorer people who depend for survival on natural resources extracted from fragile ecosystems.

The classic dilemma of social interest – the good of the community versus freedom for unrestrained individual pursuit of prosperity, and the significant influence on human development of rapid technological change – are critically discussed. Issues related to conservation of biodiversity against human needs are sensitively analysed, again contrasting the attitudes of those with ample access to resources to sustain their chosen life-styles, and the disadvantaged for whom survival depends on utilising whatever biological resources are readily available. The paper cites data from energy analyses, which estimate that 44,000 teraWatt of solar energy are absorbed in the global biochemical water cycle, whereas total world energy employed in human activities is only 10 teraWatt.

Ethical questions of sustainable development arise from biodiversity depletion: eradication of inestimable millions of organisms that offer no immediate value or utility to food systems or industrial processes. Giampetro and Bukkens conclude that unrestricted population growth is unsustainable, and neglect of resource conservation by the spread of intensive agriculture results in loss of soil fertility, excessive water consumption, atmospheric and water pollution by the products from fertiliser breakdown; that present scientific knowledge is inadequate to determine costs and benefits of many and diverse patterns of human development. Uncontrolled growth of human population activities will inevitably result in severe famines and other catastrophes.

Economic, environmental and social considerations

In a detailed series of essays, a group of authors connected with the Global Development and Environment Institute at Tufts University and headed by Jonathan M. Harris, discuss sustainable development from several perspectives [Harris *et al* 2001]. At the outset the authors discuss a growing recognition of three essential aspects of sustainable development:

a. Economic – an economically sustainable system must be able to produce goods and services on a continuing basis, to maintain manageable levels of government and external debt, and avoid extreme sectoral imbalances that damage agricultural and/or industrial production.

b. Environmental – an environmentally sustainable system must maintain a stable resource base and avoid overexploitation of non-renewable resources systems … including maintenance of biodiversity, atmospheric stability and ecosystems not always looked upon as economic resources.

c. Social – a socially sustainable system must achieve fairness in distribution and opportunity among all persons with adequate provision of such social services as health, education and gender equity.

In substantial detail the essayists discuss:

- the environmental dimension: natural capital, environmental and ecological economics, intergenerational equity in access to environmental resources;

- the social dimension: reconciliation of environment and development, governance related to provision of social services;

- imbalance between the affluent 'North' and the underprivileged 'South, taking account of unrewarded exploitation of biodiversity in Third World nations, and biased uneven confrontations between rich and poor in negotiations related to resources, trade and atmospheric pollution;

- demographics of population growth, urbanisation with particular reference to India's mega-cities, industrial spread, exploitation of natural resources;

- constraints on agricultural production and hopes for future food security, sustainable use of natural resources, potential risks posed by transgenic organisms, conservation of fisheries and forests;

- energy and climate change;

- 'globalisation' described as the accelerated integration of the world's economies through liberalisation of trade and investment; variable consequences of structural adjustment programmes among poorer nations, comparing the contradictory opinions of observers who proclaim the positive macroeconomic benefits of structural adjustment policies and their opponents convinced that social and environmental costs outweigh short term economic gains.

The authors contrast the opinions of those: a. who perceive global economic integration as adverse to the global environment and to employment opportunities, not least among workers and their unions in the affluent North; b. who regard 'globalisation' as an exceptional opportunity for expansive, economic development. The authors conclude that sustainable 'globalisation' demands planned cooperation among transnational corporations, international agencies, governments and the communities they govern who may be most affected.

Adjusted net savings

The World Bank *Development Report for 2003* presents a formula for Adjusted Net Savings as a composite indicator of national developmental progress. According to the Bank's economists, 'Adjusted net savings' equals net domestic savings (the difference between gross domestic savings and consumption of fixed capital) plus education expenditure, minus energy depletion, mineral depletion, net forest depletion and carbon dioxide damage. From this complicated formula the Bank's economists calculate the following as Adjusted Net Savings (as per cent of GDP) for groups of nations classed a. by income, and b. by region:

Low Income	7.8	High Income	13.5
East Asia & Pacific	25.2	Sub-Saharan Africa	3.9

No doubt such computations offer an intriguing exercise for highly qualified economists, but one must wonder at the reliability of the required data available from most of the poorer nations, few of which employ the capacity accurately to collect and analyse such data. The Bank report goes on to say: "While recognising the need for an aggregate index … it is important to note that indicators are most useful when they address specific problems."

The Bank report advocates the importance of 'Green Accounting'; a 'Green GDP' is claimed to be more representative of resource and environmental conservation than the classical concept of GDP, which is not a true index of the state of human welfare in any nation. Green Accounting attempts to modify national accounts to include environmental damage, environmental services and change in stocks of natural capital.

Uncertainties unresolved

The foregoing is but a sample of different persons' and agencies' conceptions of 'sustainable development'. Perhaps in common with this author after reading the foregoing some readers will be reminded of Edward Fitzgerald's translation of *The Rubaiyat of Omar Khayyam:* and consider they came out by the self-same door as in they went. It is clearly evident that 'development' and 'sustainable development' are interpreted differently by different people. Consequently, unless the purposes and objectives are accurately defined and the criteria by which 'development' will be measured or assessed are precisely stated, neither 'development' nor 'sustainable development' can be sensibly interpreted, evaluated or discussed. Since precise definitions and criteria are so frequently absent, much that is proposed for 'development' and 'sustainable development' is exceedingly difficult to comprehend. Indeed, one must question whether many of those who advocate 'sustainable development' have formulated for themselves any precise definition of purpose and objectives, of criteria by which progress can be measured, or are motivated by any systematic understanding of all that is at issue. Unless and until all who advocate and seek to implement 'sustainable development' accept and are guided by the advice of Kelvin and Lavoisier rather than by Humpty Dumpty, confusion and conflict will inevitably prevail and persist.

3

International Development:
In the Beginning

The Atlantic Charter

The meetings in Versailles following World War I, which led to the creation of the United Nations' predecessor, the League of Nations, the unwillingness of the United States Congress for its country to be a member of the League, and the manner in which the League finally collapsed are comprehensively described and discussed in a very scholarly book [MacMillan 2003]. Despite the failure of the League, the concept of an international organisation designed to ensure world peace and promote cooperation among nations did not die.

On 12 June 1942, on a battleship somewhere in the Atlantic Ocean, US President Franklin D. Roosevelt and British Prime Minister Winston Churchill conceived and promulgated the Atlantic Charter: the critical conditions to be accepted and adopted by nations to ensure peace for all humanity after the end of World War II. The Charter perceived a peaceful world in which all people would enjoy freedom from want, fear, hunger, poverty and ignorance; would have access to health services by which to eradicate chronic diseases. The Charter specified: a. the right of all people to choose the form of government under which they wished to live; b. all nations to have access on equal terms to the trade and raw materials needed for economic prosperity, and c. all nations to abandon the use of military force to pursue their ambitions or to impose their will or ideologies on other nations. It is melancholy that the present political leaders of the United States and Britain do not appear to embrace the benign vision and aspirations of their predecessors.

In 1942 representatives of the 26 nations that were fighting against Germany and Japan signed a Declaration accepting the principles of the Atlantic Charter. A year later heads of government from the United States, the United Kingdom, the Soviet Union and China agreed to create a new international organisation to pursue the principles of the Atlantic Charter. In 1944, the four nations met at Dumbarton Oaks in Washington DC and drafted a Charter for a new United Nations. The United Nations Conference on International Organisation met in April 1945 in San Francisco with delegates from 50 nations who, during the next two months, agreed on the contents of a Charter for the UN which laid down the Organisation's purpose, principles and structure. The UN officially came into existence on 24 October 1945 since when its member nations have grown to 191.

The United Nations and its Charter

The intentions prescribed in the UN Charter are noble and commendable. It is indeed sorrowful that, in their pursuit of political power and influence, several of the principal signatories to the UN Charter have repeatedly ignored the responsibilities to which they committed themselves. The Charter states: "We the peoples of the United Nations [are] determined to save succeeding generations from the scourge of war ... to reaffirm our faith in fundamental human rights, in the dignity and worth of the human person, in the equal rights of nations large and small; ... to promote social progress and better standards of life ...; to practise tolerance and live together in peace with one another ...; to unite our strength to maintain international peace and security; ... to employ international machinery for the promotion of the economic and social advancement of all peoples."

The international monetary system

In 1944, a group of nations met at Bretton Woods in New Hampshire and created the International Bank for Reconstruction and Development (the World Bank) and the International Monetary Fund (IMF) by which "the world consciously took control of the international monetary system" [Plumptre 1977]. It was decided, and has so remained ever since, that the headquarters of both IBRD and IMF would be located in Washington D.C., that the President of the World Bank would be a US citizen, and a European would preside over the IMF. In some critics' opinions this has led to the IBRD and IMF being excessively influenced by US political and economic philosophies. The recent appointment by President Bush of one of the principal architects of the war against Iraq to be President of the World Bank is particularly disturbing.

The IMF was created to assist nations that suffered balance of payment difficulties in a world economy of fixed exchange rates. It is now under pressure to monitor the relative values of national currencies, motivated by the US-perceived imbalance of trade with Asian countries, whose currencies the United States considers to be undervalued.

The UN and its Specialised Agencies

Pursuit of the peaceful and humanitarian purposes prescribed in the Atlantic Charter required a family of diverse international organisations to bring them into being, a need gradually fulfilled by the United Nations and its family of Specialised Agencies. The early history of the UN and its Specialised Agencies is described by Jones [1965]. Each of the Specialised Agencies, such as the Food and Agriculture Organisation (FAO), the World Health Organisation (WHO) and the UN

Educational, Scientific and Cultural Organisation (UNESCO), is individually chartered and supported by those national governments that choose to be its members. Not all 191 nations who are members of the UN are members of all the Specialised Agencies.

By 1965 there were 14 UN Specialised Agencies including IBRD, IMF and the International Development Association (IDA) created to act as a soft loan window of the World Bank to offer loans free of, or at low rates of interest. In some instances, IDA loans are repayable in local currencies. Four other activities related to the UN included the World Food Programme (WFP) designed to provide food aid to nations that suffered severe food insecurity; the UN Children's Fund (UNICEF); the Expanded Programme of Technical Assistance (EPTA) and the UN Special Fund (UNSF). By 1966 EPTA and UNSF were united into the UN Development Programme (UNDP) that finances development projects directly to developing nations or administered by one of the UN Specialised Agencies.

Liberation from colonialism

During the quarter-century following World War II some 60 new sovereign and independent nations, formerly ruled as colonies by Europeans, came into existence. The economic, material, skilled and educated human resources varied considerably among the newly emerging nations. All, in varying degrees, had – and many still have – need of economic and technical assistance, at first provided by the UN agencies, subsequently by governments of the more affluent North American, European and Oceanic nations. The needs, capabilities and resources among the new sovereign nations differed enormously, in part attributable to how their boundaries were defined, how they were colonised, and the sensitivity of their former colonial masters, some of whom bequeathed institutional facilities for education, agricultural and economic development, others left a legacy of massive illiteracy.

Colonisation of Africa

In 1884 Otto von Bismarck, then Chancellor of Germany and the principal architect of German unification, convened a conference in Berlin, ostensibly to set a pattern for international economic development of Africa. Nations who attended included the United States, Austria, Britain, Belgium, Denmark, France, Germany, the Netherlands, Norway, Italy, Spain, Sweden and Turkey. The effective outcome was for Africa to be carved up into a series of colonies with little regard for existing geographical distributions of indigenous tribes and ethnicities. Russell Warren Howe [1967] cogently describes the outcome: "Leopold, King of Belgium, was free to build his Congo Free State [and] the inevitable principle of conquest and colonisation was applied across the thousands of miles between the Cape and Cairo". Bismarck stressed the need: "to associate the natives of Africa with civilisation by opening up the continent to commerce". Yet no person native to Africa was invited to the Berlin Conference, nor were Africans invited to comment on the outcome [Howe 1967]. Kennedy [1988] describes how many of the affluent nations acquired and eventually gave up their African colonies.

Bismarck's opening up of the African continent to commerce resulted primarily in the African colonies being a source of valuable raw materials to be profitably processed and sold by their colonial masters' industries. Though the present pitiful state of several sub-Saharan African countries is attributable to indigenous government mismanagement and corruption, the root cause of inter-ethnic and inter-tribal conflict had its origins in the ethnically illogical carving up of African territories by Bismarck's Berlin conference.

Conflict in the Middle East

The pursuit of profit by devious acquisition of subjugated nations' resources, evident across Africa, is equally responsible for the existing dangerously disastrous conditions in the Middle East and West Asia.

The prime purpose of the deception practised on the Arab nations by Sir Arthur Balfour and Sir Mark Sykes of Britain, and George Picot of France, later ratified by these nations and the United States in the Treaty of Versailles, was then and remains today, simply to gain control of the region's oil reserves and the nations that own them [MacMillan 2003]. Western governments and their allies, militarily active in the Middle East and West Asia, ignore the Atlantic Charter provisions that all nations have the right to choose the form of government by which they will be ruled, and that no nation has a right to impose its ideology or concept of government by military force.

International development conferences

The second half of the twentieth century brought forth countless conferences, missions and publications about 'International Development', in principle deontological: the duty of affluent nations to improve the quality of life of less fortunate people; in practice more often motivated by donors' political priorities. The following are some of the important conferences, where they were convened and what resulted:

1944	Bretton Woods	Creation of IBRD and IMF
1945	San Francisco	Creation of United Nations Organisation
1945	Quebec City	Foundation of FAO
1960	Washington	Creation of International Development Association (IDA)
1971	Washington	Founding of Consultative Group on International Agricultural Research
1972	Stockholm	UN Conference on Environment (led to UN Environment Programme)
1975	Rome	World Food Congress (established World Food Council – later disbanded, and the International Fund for Agricultural Development – IFAD)
1977	Nairobi	International Conference on Desertification

1978	Singapore	International Conference on Science and Technology for Development
1981	Nairobi	World Conference on Energy
1990	New York	World Conference on the Child
1992	Rio de Janeiro	UN Conference on Environment and Development
1993	Vienna	UN Conference on Human Rights
1994	Cairo	UN Conference on Population and Development
1995	Copenhagen	UN Conference on Poverty and Social Development
1995	Beijing	World Conference on Women
1995	Rome	World Food Summit
1997	Istanbul	World Conference on Urban Quality of Life
2002	Johannesburg	UN Conference on Environment and Development (Review after 1992 Rio)
2004	Bangkok	One of a series of conferences on HIV/AIDS
2006	Toronto	Most recent HIV/AIDS conference (30,000 delegates)

In relation to the last two in the list, can there be doubt that governments and those who seriously care are unaware of the present and future potential human devastations attributable to the HIV/AIDS endemic? Is there any persuasive need for more international conferences to publicise the horrors of AIDS? How many doses of antiviral drugs and how much medical aid for AIDS/HIV victims could have been provided by the millions of dollars spent to transport and accommodate delegates at huge international conferences? It is this author's opinion that most of these mammoth international conferences serve only to provide a forum for politicians and speechmakers to present pious pronouncements on the conference subject. The inevitable demands for substantial essential investments and actions are rarely provided.

Systematic evaluations of the costs of convening huge international conferences versus specific tangible benefits derived are difficult to discover. The costs are invariably high, given that most of the senior politicians and bureaucrats travel first class and are accommodated in the most expensive hotels. It is interesting to speculate how many poor people could have been provided with clean water, improved sanitation, access to primary education with the money spent to convene the 1992 and 2002 conferences on environment and development, convened, respectively, in Rio and Johannesburg.

Purposes of international development programmes

The opening sentence of *Partners in Development*, the Report of the 1968 Pearson Commission on International Development, states: "The widening gap between the developed and developing countries has become a central issue of our time" [Pearson `1969]. In the words of the familiar French *dictum*: "Plus ça change, plus c'est la même chose". Despite the wise advice of Cicero, Kierkegaard, Churchill, Santayana and others, that those who fail to learn from history are prone to repeat the mistakes of the past, development agencies and politicians repeatedly act as if unaware of, or unwilling to be influenced by, past experience. Few are the development agencies that maintain a systematic corporate or institutional memory. The penchant to latch on to transitory fashionable slogans, for which no precise definition is stated, is all too frequent.

Development mistakes, misfortunes and progress

A report on development issues by a Canadian parliamentary committee quotes Brian Walker, a former executive director of Oxfam: "The field of development is a veritable junkyard of abandoned models, each focused on a particular aspect while ignoring the rest." [Winegard 1987]. Walker goes on to say: "The longer I work in the field of aid and development, the more I appreciate the complexities of the process." Expressing a similar sentiment in the agency's 1985 Annual Report, Margaret Catley-Carlson, president of the Canadian International Development Agency, wrote: "Is the developing world really developing? My judgment leaves me full of doubt, development is a long uncertain process."

These observations illustrate the indisposition of development agencies to conduct rigorous, critical systems analyses from which to formulate their priorities and programmes. Equally unfortunate are the many efforts to transfer technologies with little prior systematic study of the resources, opportunities and constraints, of the social, economic and physical environments of the intended recipients. Though significant progress has been realised among nations variously classed in the 1960s as 'low income', 'undeveloped', 'less developed' or 'developing', much effort and expense has been wasted on ill-conceived projects by agencies seemingly unaware of modern systems analysis.

It is not unusual among newcomers to development assistance to seek simple solutions to highly complex conditions and difficulties, to attempt transfers of isolated technologies and processes between widely different communities. India is all too familiar with foreign 'experts' who arrive with solutions seeking problems, innocent of many centuries of Indian experience. The Pearson Report [1969] stated: "The international aid system, with its profusion of agencies, lacks direction and coherence." Little subsequently seems to have changed.

Since the 1960s, India and other Asian nations have made impressive progress in agricultural, economic, industrial and technological development. To what extent progress is attributable to external assistance or to indigenous effort defies determination. By conventional economic standards, India has progressed faster than industrialised countries during their early stages of development. Between 1790 and 1820, the period of the so-called industrial revolution, Britain's Gross Domestic Product is estimated to have grown at about 2 per cent per year. Between 1990 and 1999, India's average annual growth in GDP was above 6.0 per cent. While recognising that the scientific and information resources available during the industrial revolution were fewer and less advanced than those accessible to India, India's economic progress is, nonetheless, a tribute to the wisdom, since independence, of India's sustained investment in research, development and higher education. Since the 1960s Official Development Assistance (ODA) by OECD nations to poorer nations has been anything but sustained.

Table 3a
ODA from OECD Nations

	1965	1970	1975	1980	1985	1990	1995	2000	2002
OECD Total									
$Bn USD	6.5	7.0	13.9	27.0	29.4	53.4	54.5	52.3	58.3
OECD % GNP	0.48	0.34	0.35	0.37	0.36	0.34	0.30	0.24	0.23
USA % GNP	0.49	0.31	0.27	0.25	0.24	0.15	0.13	0.11	0.13
UK % GNP	0.47	0.37	0.37	0.35	0.33	0.31	0.32	0.32	0.31
Canada % GNP	0.19	0.37	0.54	0.52	0.49	0.47	0.45	0.22	0.28
Denmark % GNP									0.96
Norway % GNP									0.89
Sweden % GNP									0.83
Netherlands % GNP									0.81

Only the Scandinavian nations and the Netherlands have realised the recommended target of 0.7 per cent of GNP. What is classed as ODA and to whom assistance is provided varies among donors. Some donors include gifts of armaments in their ODA calculations. The United States, lowest among OECD nations as per cent of GNP, gives most of its aid to a Middle Eastern nation, which by any accepted international standard cannot be considered poor or underdeveloped. While close to 40 per cent of Scandinavian ODA is given to poorest nations, only 15 per cent of US ODA is so directed.

Table 3b
UNDP estimates of ODA flows in 2004

	Total (US$bn)	per cap (US$)
World	69.8	
Low-income countries	32.1	13.7
Middle income	19.0	8.4
All developing nations	65.4	9.7
Sub-Saharan Africa	22.7	32.9
South Asia	6.6	4.3
Latin America/Caribbean	6.1	9.9

During 2004 more than US$ 400 billion was invested in waging war and reconstruction projects in Afghanistan and Iraq, an amount forecast to rise significantly in the future. These are investments that progressively deplete ODA to the poorest nations. In the 8 December 2006 edition of the *Guardian Weekly* Julian Borger describes how vast sums intended for construction in Iraq are squandered by corruption and companies who overcharge for their contracted services.

Rich nations and poor nations

Ward [1961] and Galbraith [1965] drew attention to inequitable income and resource distribution both between rich nations and poor nations, and within national jurisdictions significantly variant in political ideologies. Gross inequities between the rich and poor exist within both affluent and less affluent nations. While poverty among the poorest nations may in some degree be alleviated by constructive development assistance, inequities between rich and poor within nations must be mitigated by those who govern them.

Modern development economics, which originated during the 1940s, relates to countries or regions characterised as under– or less developed relative to the rest of the world. Earliest objectives, variously described by Pearson, Brandt and Brundtland, were to raise standards of living throughout the poorer nations by providing more essential goods and services to fast expanding populations. The International Monetary Fund, the World Bank and the General Agreement on Tariffs and Trade, which came into being in the late 1940s, were intended to provide a stable framework for economic and social development among all nations. Criteria by which to assess human development diversified from simple estimates of GNP or GDP/Cap to include such basic needs as education, nutrition, health services and status, hygienic sanitation and employment opportunities particularly for rural and urban poor.

Several observers propose drastic reforms for the IMF and World Bank and some UN agencies. Whether the United Nations as originally conceived should predominantly serve the interests and satisfy the needs of the poorest nations, or become the instrument whereby the most powerful impose their ideologies on the rest of the world, is at the root of the dispute over UN reform. There are many who contend that the UN and its agencies should exert greater influence on nations' policies and courses of action that are detrimental to the environment and tend to exacerbate conflict. Of particular concern and cause for disagreement is which nations should hold permanent membership of the UN Security Council and what powers should it exercise. The present restriction to five members: China, France, Russia, Britain and the United States is clearly unrepresentative of the membership as a whole. The powers of the Security Council are repeatedly constrained by the right of any one of the five to veto decisions. The need for reform of the United Nations and its degree of authority over international issues will probably remain matters of dispute and disagreement for some time to come.

While recognising that non-government organisations (NGOs) are widely divergent and variable in resources, capabilities, experience and ideologies, several have provided a valuable counterbalance to World Bank, IMF and other international agencies' programmes. Various NGOs assess the effects of development programmes on primary health care, safe drinking water, biodiversity protection, pollution control and overall ecological conservation and environmental change. They draw attention to development activities detrimental to environmental degradation, social inequity and political mismanagement and corrupt diversion of resources provided [Fox & Brown 1998]. While recognising that NGOs possess hugely different resources and capabilities, some have marshalled substantial resources and acquired exceptional experience. It seems unfortunate that only 0.02 per cent of OECD ODA is channelled through NGOs.

Assessments of international development programmes

Most investment in international development programmes, whether multilateral – channelled through United Nations and related agencies, or bilateral – direct assistance from donor to recipient nations, comes from government treasuries and their taxpayers. During the past four decades, a bewildering sequence of conferences, commissions, committees and consultants have been convened and contracted to assess progress in development among nations originally designated as 'developing' or 'third world'. It is a recent fashion to describe such evaluations of development as 'Impact Assessments', a term as semantically indefinable as 'sustainability'. 'Impact' is derived from '*impactus*' the past participle of the Latin verb '*impingere*' meaning: 'to attach to' or 'to press two objects close together'. During World War I, 'impact' was adopted by the British army as a military term to describe the result of particular missile strikes or explosions. 'Impact assessment' literally assumes a single dramatic event; it is semantically meaningless when applied to the evaluation of a diverse complexity of economic, environmental, social and technological interactions variously constituent to development programmes and processes. To speak of 'negative impact' is a semantic nonsense.

A particular difficulty inherent in assessing progress in development programmes is that those who designed and started them, failed to define specific determinable and quantifiable objectives and criteria. During the early years, substantial investments were made on the naïve assumption that to create institutions, provide equipment and foreign expertise would inevitably result in progressive development. Missing were precise definitions of objectives, perceived benefits to be realised, by whom expected benefits were to be realised, what were the potential constraints, risks and adversities that might be encountered. Not unusual was a concept of 'one solution satisfies all needs'. During the late 1960s, this author encountered 'an expert' from a UN agency who presented an identical plan for agro-industrial development for Iran and Uganda, countries vastly different in traditions, cultures, ecologies, climates and resources.

Nonetheless, from many published reports, a vast volume of data, descriptive material and observations relevant to development have been recorded. Whether assessed by classical economic criteria, industrial or social indicators, progress among nations that have received development assistance is highly heterogeneous. At the time they became independent self-governing states in the 1960s, Ghana and Nigeria were more prosperous than the Republic of South Korea. According to recent World Bank data, GNP/Cap in South Korea is now *ca* $8,500 USD, in Ghana $390, in Nigeria $280. Expressed in Purchasing Power Parity, GNP/Cap in South Korea, Ghana and Nigeria are, respectively $11,500, $1,990, and $1,220. While the financial and industrial economies of many once poor Asian nations have expanded impressively, countries in sub-Saharan Africa have stagnated or, in several instances, retrogressed. During the 1980s, Zimbabwe was among the most productive of African nations: being a net exporter of cereal grains, with a well-developed agribusiness sector, and a thriving university. Zimbabwe's economy, agriculture and agribusiness are now in serious decline. Between 1990 and 1995, Zimbabwe's food production fell by over 30 per cent, investments in the military exceeded support for health and education. Since 1980 the exchange rate of the Zimbabwe dollar against the US dollar has dropped from near parity to US$ 1.0 = Zim$ 100,000. The rate of inflation in Zimbabwe now exceeds 1400 per cent per year and is predicted to rise even higher.

A detailed record of the findings and recommendations of the many extensive and expensive assessments of development would fill countless CD ROMS. What follows is a brief summary of some of the more prominent, including the series of annual World Bank Development Reports; the Pearson, Brandt and Brundtland Commissions, convened and financed by the World Bank; two international conferences on environment and development, the first in Rio in 1992, and the second in Johannesburg in 2002.

World Bank Development Reports

The series began in 1978, each report addressing a particular development issue, with statistical analyses, commentaries and recommendations. The reports are compiled by World Bank staff and include a disclaimer that judgments and opinions are those of the contributors and not necessarily reflective of the Bank's Board or Executive Directors. The authors do not guarantee the accuracy of the data presented, a not surprising caveat given that statistical tables include estimates of total populations, GDP or GNP/Cap, macroeconomic indicators, income distribution, data related to health and education, commercial energy generation and consumption, land use and rate of urban growth, external debt, value of exports and imports. Analyses cover more than 125 nations. The cost and complexity of carrying out a reliable population census or determination of income distribution are considerable in any jurisdiction. In nations where frequent migrations are exacerbated by conflict, civil disturbance, food and other insecurities, where the public service lacks essential resources and competence, the collection and maintenance of accurate statistical records is virtually impossible. The World Bank data is therefore, to be regarded as broadly indicative rather than precise.

The following are subjects addressed in World Bank Development Reports (WDRs) from 1978 to 2004.

1978 Review of development progress 1950–1973
1979 Development prospects and international policies
1980 Poverty and human development
1981 Prospects for international trade
1982 Agriculture and economic development
1983 World economic recession; Prospects for recovery
1984 Population change and development
1985 International capital and economic development
1986 Prospects for sustained growth
1987 Industrialisation and foreign trade
1988 Managing the world economy
1989 Financial systems and development
1990 Poverty
1991 The challenge of development
1992 Environment degradation
1993 Investing in health

1994   Infrastructures for development
1995   Workers in an integrating world
1996   From plan to market
1997   The State in a changing world
1998   Knowledge for development
1999   Entering the twenty-first century
2000   Attacking poverty
2002   Building institutions for markets
2003   Sustainable development
2004   Making services work for people
2006   Equity and development

The World Bank Development Reports provide the most continuous source of comprehensive information and data on 'development', broadly defined, covering most of the world's nations. The text and commentary of each report covers close to 200 pages with up to 20 statistical tabulations. The abbreviated content and commentaries of only a selected few will be summarised.

WDR 1978

The first report, published in 1978, reviews patterns and progress in development from 1950 to 1975. It states that, in general, conditions among low-income developing countries had improved significantly since 1950. Many of the LICs had shaken off their colonial shackles and were now governed by indigenous rulers. The Bank estimated that average incomes per capita across all LICs had risen by 3.0 per cent/ year, life expectancy at birth from 36 to 44 years [cf. 72 yrs for OECD nations], infant mortality fell from 142 to 122 per 1000 live births [cf. 15 in OECD].

Nonetheless, Robert McNamara, then World Bank President, stated that over 800 million, most inhabitants of rural ecologies in sub-Saharan Africa and Asia, were "trapped in absolute poverty", characterised by chronic malnutrition; illiteracy; disease; squalid, unsanitary dwellings;

high infant mortality and low life expectancy. Agricultural production, on which most LICs economies depended, had increased on average at 2.1 per cent/year. Total LIC populations were however increasing at 2.4 per cent year and had almost doubled between 1950 and 1975. An added constraint to food security: urban populations in Sub Saharan Africa, Latin America and Asia were growing, respectively, at 5 per cent, 4.3 per cent and 4.0 per cent/yr.

The report noted that total ODA increased more than four-fold between 1960 and 1978. Expressed as a proportion of donors' GDP, contributions from the United States fell from 0.53 to 0.26 per cent, while Canada's ODA rose from 0.19 to 0.54 per cent. In its forecast of future population growth, the Bank underestimated the LIC population in the year 2000 by over 500 million. WDR 1978 contended that progress in improving living standards was slow and uneven; that the highest priority for development programmes should be to alleviate poverty by accelerating economic growth in the LICs. Specifically, it recommended policies and actions to raise employment opportunities and income among the poorest rural communities, who should be allowed greater access to essential educational, health and social services. The report drew attention to gross international disparities in consumption of commercial energy. In 1975, the industrial nations consumed an average of 3500 kg of oil equivalent [Kg OE] *per caput*, in contrast to the LICs consumption of 35 Kg OE/cap.

WDR 1985

The report entitled *International Capital and Economic Development*, states that from 1950 to 1980 average annual income growth was estimated as 3.1 per cent/year for industrial nations, 3.0 per cent for Middle Income Countries [MICs] but only 1.3 per cent for LICs. Authors suggest the advantages and risks of capital flows from richer to poorer nations. Recipients that used imported capital prudently enjoyed accelerated economic and productive growth while imprudent recipients, particularly those who spent foreign capital on armaments, slowed their economic growth and raised their foreign debt. Between 1970 and 1985 the total accumulated medium and long term debt among developing countries increased from *ca* $70 billion to *ca* $700bn USD; among poor African nations, total debt rose from 19 to 58 per cent of GNP. In addition to being seduced into military adventures, during the late 1970s and early 1980s, poorer nations' economies were heavily stressed by increases in the price of oil, rising interest rates, subsidised exports and other impediments to free and fair trade by European, North American and other affluent nations.

WDR 1990

This report devoted to *Poverty* estimated that over one billion people in the LICs suffered abject poverty, most being inhabitants of South Asia and sub-Saharan Africa. With 30 per cent of total world population, South Asia was home to 50 per cent of the world's poorest folk. In South Asia 520 million, in sub-Saharan Africa 180 million survived on incomes of less than $2.00 USD per day (by the year 2000 these numbers, respectively, had risen to 1.2 billion and 570 million). To alleviate the deplorable incidence of poverty, the Bank recommended policies and investments to encourage labour-intensive employment particularly among rural communities. Effective expansion of rural employment required substantial investment in rural infrastructures, transport and communications, reliable electrical energy, clean water, technical and financial advisory services and access to markets. Also urgently needed were improved access to rural health and social services.

WDR 2000

The 2000–2001 report is mainly devoted to poverty alleviation, which is discussed in a later chapter. While expressing optimism that donor nations have renewed and re-energised their commitments to development and poverty alleviation, this optimism is not supported by recent declines in aid from OECD nations expressed as a per cent of

GNP. For some donors, aid is a means of imposing their ideologies or of deriving economic benefit to their economies and industries: such being the case with most tied aid. The Report deplores a lack of coordination and focused priorities among donors, with too little consultation between donors and the nations and communities to whom aid is directed. During the 1990s some 40 donors supported over 2,000 different projects in Tanzania. During the same period, 65 distinct government and quasi-government agencies in Ghana were receiving aid. It is not unknown for governments in poor nations to use aid to replace rather than to supplement government investments in agriculture, health, education, social and other essential services. WDR 2000 states that donors are inclined to dominate recipients and do not consult sympathetically with those that projects purport to assist. Before embarking on new projects, donors take too little time to understand local conditions, resources and constraints. They seem unaware of the advice propagated by a North American stockbroker: 'Investigate before you invest.' Unsystematic planning and failure to define precise objectives proscribe efficient monitoring and assessment of progress. (It is this author's experience that few donors seem to monitor and assess what transpired in consequence of their projects, several years after the projects ended. Indeed, all too frequently donors appear to lose interest the moment a project comes to an end).

WDR 2000 restates Pearson, Brandt and Brundtland, that there are no simple strategies or rapid solutions to the complexities of poverty and underdevelopment. Populations among some donor nations appear to believe that because of mismanagement much foreign aid has neither reached nor benefited those for whom it was intended. WDR 2000 again urges donors to abandon tied aid, to contribute more to debt relief, to channel more aid monies through demonstrably capable NGOs. It estimates that less than 10% of government aid is administered by NGOs (as noted above, this is a generous estimate). Donors should employ people of relevant experience, people who can judge wisely the needs and opportunities of recipients; who are able to monitor progress, to decide when changes in activities are needed and, most difficult, when a project should be terminated. Project design, development and monitoring cannot be productively brought about by people who make short infrequent visits from a donor's headquarters several thousand kilometres away from where the action is.

WDR 2003

WDR 2003, entitled *Sustainable development in a dynamic world*, begins by asking: "How can productive work and a good quality of life be provided for almost 3 billion people who subsist on less than $2.00/ day, and the 3 billion forecast to be added over the next 50 years ... in an environmentally and socially sustainable way during transition to a predominantly urban world?" During the next 30 years, among developing nations, while rural populations are predicted to remain stable or contract, urban populations will probably double in magnitude; mega-cities, each with a population in excess of 10 million, will increase from the present 16 to over 30 by 2025. Particular attention must be directed to many millions who in abject poverty struggle to survive on fragile lands, in arid ecologies with uncertain rainfall, on soils low in fertility and subject to persistent erosion. WDR 2003 states during the next 50 years, LICs must achieve substantial growth in income and productivity. There is need for shrewd management of social, economic and environmental issues and opportunities during an inevitable transition to an urban society. Development objectives must give priority to elimination of rural and urban poverty; intensification of agricultural production; sustainable management of land and water to provide food adequate for expanding urban populations; improved efficiency of food protection, preservation and distribution from rural producers to urban markets. Equally urgent is the need to provide off-farm employment in rural agribusiness.

In dry regions water scarcity will become the most critical constraint to agricultural, economic and social development. Domestic, agricultural and industrial water demand cannot be satisfied simply by water reallocation or structures to distribute water. Agricultural production must be more conservative of water; agriculture, the largest consumer of water, must give some way to other demands. Provision of adequate water demands substantial investment in desalinisation, in purification and recycling of waste water.

Cities will expand and their populations become more congested because of rural-to-urban migration, by reproductive increase among present inhabitants, and through spread of domestic dwellings and industries into rural areas. Unplanned, uncontrolled urban expansion leads to slum proliferation; inadequate safe potable water, hygienic sanitation and waste disposal; disease epidemics from over-crowding; water pollution; malnutrition and inadequate health care; poor facilities for public, private and commercial transportation; and rising atmospheric pollution from industrial and motor vehicle exhaust emissions. Air and water pollution, disease epidemics and human migrations that cross national borders are of long painful experience. Regrettable indeed are the governments who selfishly refuse to comply with international environmental protocols or to accept censure for their excessive pollutions, for failure to conserve and protect their and their neighbours' environments.

In planning for sustainable development, WDR 2003 proposes a time horizon of 20 to 50 years over which to predict probable environmental and social consequences of development alternatives. The foremost challenge to development agencies and governments is to alleviate extreme poverty while protecting the environment and conserving critical resources. Over the past 30 years, while world population rose by 2 billion, among developing countries average *per capita* annual income, in constant 1995 dollars, increased from US$990 to $1,350; infant mortality was cut in half from 107 to 58 per 1000 live births, adult illiteracy fell from 47 to 25 per cent. A drop in the numbers who subsist on less than $1.00/day can be attributed to exceptional economic growth in the Peoples Republic of China. In contrast, throughout sub-Saharan Africa, abject poverty persists and is worsening among many communities. While overall economic growth among SE Asian nations has been impressive, two-thirds of the world's poorest people live in East and South Asia.

WDR 2003 itemises various assets essential to development: human assets – the skills, talents, and abilities of individuals and institutions as influenced by educational and training facilities; renewable and non-renewable natural assets such as forests, fisheries, mineral deposits, air, water and soil; created assets including machinery, equipment, structures and financial resources. Social amenities and services are difficult to quantify particularly when strongly influenced by local customs, cultures and traditions.

WDR 2003 indicates the institutions and organisations that must cooperatively contribute to sustainable development: government agencies, commercial industries, civil society institutions, environmental and political organisations. It advocates institutional reform to ensure that what may appear as desirable development does not entail social and/or environmental instability. It suggests how livelihoods on fragile lands might be improved; the urgent need to eradicate rural poverty, to strengthen rural-urban integration, to intensify agricultural production, prudently and conservatively to manage land and water; the need to devise efficient integrated food systems to ensure urban food security together with rural prosperity. The report deplores how excessive subsidies to agriculture among affluent nations have depressed world food and commodity prices to the disadvantage of farmers and poorer nations' economies. While food at affordable prices will be abundantly accessible to affluent families, given present trends future food security among poorer folk cannot be assured. Rural producers will need to increase their crop and livestock production both to satisfy their own needs, and for sale to local and distant markets.

The report illustrates interactions among access to agricultural land, resource degradation and rural poverty. Over the past 30 years among developing regions land degradation has resulted in a drop of 12 per cent in crop production. Studies of cropping patterns and fertiliser use in Africa reveal that failure to restore soil nutrients is a serious impediment to crop production and is both a cause and result of widespread poverty.

Though by global macroscopic assessments land and water may appear abundant, objective disaggregated analyses indicate regional deficiencies and overexploitations on all continents, the most serious being in sub-Saharan Africa. The World Commission on Water predicts demand will rise by 50 per cent over the next 30 years and that 4 billion people, roughly half the world's population, will suffer water stress by 2025. Demand for scarce land and water is a potential cause of military conflict in the Middle East and North Africa where deficiencies and consequent competition are most evident.

WDR 2003 speaks of externalities where the actions of one nation or community can inflict environmental, economic and/or social costs on other nations, costs the guilty parties do not bear. Dams and barrages to generate electricity in an upstream country can deny farmers in downstream nations' access to water essential for crop irrigation. Acidic atmospheric pollution by one nation's industries can be carried to poison the air and inland waters of neighbouring countries.

The report comments on the evident patterns and predictable consequences of urbanisation, urging that sustainable urban development requires more careful planning and control than has been exercised in the past. While local jurisdictions must take responsibility for planning and control within their territories of responsibility, there is urgent need for recognised international organisations and authorities to monitor and coordinate conditions, and enforce regulations to prevent excessive exploitations, ecological and environmental degradation, processes and policies that contribute to climate change, human migrations and the spread of diseases that are international and transcend national boundaries.

It is questionable if the Bank's advocacy of the Marshall Plan for the rehabilitation of post-war Europe is an adaptable example for programmes of assistance to poor African nations. Though much of Europe's industrial and productive capacity was devastated by military actions during World War II, post-war Europe was home to many scientists, technologists and administrators with long experience of agricultural, industrial and governmental management. Human and other resources available to poorer nations are incomparable with those that existed in post-war Europe.

In its closing summary WDR 2003 presents a weary litany of oft-repeated ills and shortcomings in present and past development programmes, of earlier recommendations consistently ignored. It speaks of low productivity among LICs, home to one-fifth of the world's population who survive on less than $1.00/day, gross inequalities within and among nations: the average personal income in the 20 most affluent nations being 37 times that in the poorest, a ratio that has doubled since 1970. It deplores egregious expenditures on armaments, excessive environmental damage and stress, and the miserly response to these and other ills by those with wealth and power. It wisely states there is no simple definition of or pathway to international sustainable development, that alleviation of the poverty, ill-health and social ills suffered by so many is the responsibility of all nations, all governments, indeed of all people. It refers to the conflicting concepts of what constitutes over-consumption, and excessive exploitation of the Planet's resources between (1) environmental conservationists and (2) the most affluent governments and their industries. WDR 2003 indicates the potential devastations and degradations if all nations, in the future, were to adopt levels of resource consumption equivalent to those of the most affluent societies.

WDR 2006

Entitled "Equity and Development", *WDR 2006* discusses diverse aspects of inequity: inequities within and among nations and communities, in access to education, health services, to markets and investment, to land, equal rights and justice. (It makes no mention of inequitable treatment extended to different nations by the most powerful; that those nations considered friendly allies and of like political persuasion are treated significantly more favourably that those democratically elected governments whose political ideologies differ from the powerful).

In marked contrast to earlier *WDRs*, *WDR 2006* includes only four tabulations of development indicators:

a. Key indicators: population, GNI (gross and PPP), life expectancy at birth, and CO_2 emissions/cap
b. Millenium development goals
c. Economic activity: GDP, agricultural production, gross capital formation
d. Trade, aid and finance.

Tabulation no 4 emphasises merchandise trade and Foreign Direct Investment (FDI).

In 2003 total world FDI is estimated as $573 billion; in low-income countries *ca* $18.2 billion; in East Asia and the Pacific region $59.6 billion (herein lies a contradiction: FDI in China is quoted as $53.5 billion, in Hong Kong as $13.6 billion).

Between 2002 and 2004 net aid flows increased with a high proportion allocated to debt forgiveness, emergency and disaster relief, and to reconstruction of property and services destroyed or impaired by the military attacks on Afghanistan and Iraq. Over this period, ODA to the most highly indebted poor countries in real terms declined. In 2003, among all OECD-DAC nations ODA represented barely 0.25 per cent of GDP (in the United States it was 0.15 per cent). In sharp contrast, subsidies to their own agricultural exports was at least 1.5 per cent of OECD-DAC GDP; i.e., subsidies to agricultural exports were six times the value of OECD-DAC ODA to poorer nations.

In his foreword to *WDR 2006* the World Bank President Wolfowitz stresses the need to improve the investment climate for every nation. *WDR 2006* is relatively silent on the inequitable investment opportunities between the rich and powerful and the poorest nations. A former chief economist at the World Bank, Joseph Stiglitz, wisely observes that poor countries need much stronger economic foundations and economic aid before their markets can be thrown open to international competition.

The *WDR's* Overview states that "… economic and political inequality … lead to economic institutions that systematically favour those with more influence …" and "distribution of wealth is closely associated with social distinctions that stratify people, communities and nations into those that dominate and those that are dominated." Having held senior positions in the Republican government of the United States, President Wolfowitz is no stranger to poorer, weaker nations being dominated by financial, political and military persuasion inflicted by the more powerful, an issue so evident in the Israel-Palestinian conflict.

The Epilogue to *WDR 2006* proclaims the superiority of markets over central (government) planning and control. It is well recognised that government and quasi-government agencies frequently lack the commercial competence and experience to manage profitable business. Nonetheless, even the most powerful governments, wholly dedicated to free enterprise, use their taxpayers' money to bale out near-bankrupt airlines, to subsidise their agricultural exports to the disadvantage of poorer nations, to favour selected industries by means of expensive contracts for armaments. Totally free, unconstrained access to all foreign markets would result in absolute domination of the poorest by the most affluent, powerful nations and their favoured transnational corporations.

Nor can there be optimistic hope for equity within industrialised nations when the salaries and financial privileges granted to senior corporate executives are so vastly greater than wages paid to the men and women who labour on the factory floor or to immigrant workers.

That support for human development is critical and central to sustainable development is beyond dispute. As illustrated in the Indian case-studies (Chapter 11), poor rural women and young people can learn relatively complex skills and abilities: to own and manage profitable small business, to process locally available biological materials into preserved foods and medicines. But those who have acquired such skills must have ready access to technical and financial advisory services, best provided by more highly trained members of their own communities. Chronic poverty, with consequent malnutrition and debility, can be permanently alleviated only by enabling poor people to own, operate and be employed in profitable local industries and commercial enterprises.

The *WDR 2006* recommendation that donor-recipient relations be improved and designed to enable those, for whom aid is intended, to determine their own development priorities, policies and programmes was advocated by Pearson, Brandt and Brundtland, and has been repeated many times since. But so long as donors use aid primarily to support their own farmers and industries, the inequities of the rich becoming richer and the poor poorer will inevitably persist.

Comment on World Bank Development Reports

Each of the earlier Bank Reports (prior to *WDR 2006*) provides an extensive and timely source of statistical data and critical assessments of criteria of development among the poorer nations. While one cannot contradict many of the comprehensive recommendations in the World Bank reports, without dramatic change in present political priorities among the world's most powerful and affluent nations, there seems little to indicate that the gross inequities so widely evident between rich and poor and the immense resultant suffering will be alleviated in the foreseeable future. Given the persistent decline in overseas development assistance (ODA) to the poorest of humanity among most OECD nations, and little evidence of a unified or coherent set of development priorities, it is difficult to be optimistic that the widening gap between rich and poor within and among nations will be remedied any time soon. While economic and technical progress in Asia, particularly in India and China, is gratifying, the melancholy state of most of sub-Saharan Africa is cause for deep concern. The depressed state of sub-Saharan Africa will not be ameliorated simply by foreign investment nor by the disorganised, unfocused and uncoordinated multifarious tied-aid programmes imposed by many agencies.

It must be recognised that many of Africa's woes have their origins in the witless colonisation that emanated from the Bismarck conference, particularly in the Democratic Republic of the Congo where very young children are forced to work long hours in copper mines owned by foreign mining companies. At the same time, more than one African nation has been and is being exploited by its own corrupt rulers, in some instances kept in power by foreign nations and companies which benefit from cheap access to their mineral wealth, few of the proceeds from which bring economic or social benefit to the poor inhabitants and oppressed children.

Despite warnings from Barbara Ward and the Pearson Commission more than 30 years ago, abject poverty persists as an international social disgrace, with inequalities between rich and poor, both within and among nations, becoming ever greater. The World Bank relates how 23 per cent of all crop land, pasture, forests and woodlands have been degraded since the 1950s, a sizable proportion so despoiled as to be irreversible. Since 1960, one-fifth of all tropical forests have been decimated; between 1980 and 1995, 200 million hectares of forest in developing regions disappeared. According to the World Resources Institute [WRI 1997] barely one-fifth of the Planet's original natural forest has survived. Loss of biodiversity is a concern expressed by many sensitive observers as is the predation of indigenous species by introduction of aggressive exotics.

Assurance of economic and food security for an expanding world population, over half of whom will live in urban communities, demands international cooperation in longer term planning, a commitment by all to use the Planet's resources more conservatively and equitably. Governments must work to bring about effective integration of food and agricultural production with post-production systems: the protection, preservation, processing and distribution of foods after harvest. On a more optimistic note, information can now be rapidly transmitted electronically in immense volume, enabling countries to learn from one another and enjoy access to a vast diversity of useful information and data.

The World Bank repeatedly poses the question, "What is meant by sustainable development, and how can progress be measured?" It suggests indicators of progress are most useful when disaggregated to diagnose and analyse specific difficulties and activities. The Bank Reports speak to the long-term difficulties of identifying substitutes for resources or assets that are overstressed of over utilised. They repeat the advice of earlier soothsayers that a. development, however defined, is a dynamic not a steady state process; and b. as warned by Harry Dexter White in 1944 at Bretton Woods, all development programmes must be planned and proceed cautiously, devoting as much attention to potential risks as to hoped-for benefits.

UNDP Human Development Reports

HDR 2003

In addition to a presentation of HDI rankings, together with various data illustrative of nations' expenditures, patterns of development and states of well-being, HDR 2003 lists the stated objectives of the UN Millennium Declaration adopted in the year 2000 by 189 nations. The specific statistical objectives (e.g., reduction in poverty and hunger) are intended to be realised between 1990 and 2015:

a. to halve the number who survive on less than $1.00 per day
b. to halve the number who suffer from chronic hunger
c. to reduce infant mortality by two-thirds
d. to halve the number who lack access to safe and clean drinking water
e. to achieve universal primary education
f. to promote equality between men and women
g. to ensure global environmental sustainability.

Other objectives propose: a. improvement in the lives of slum dwellers; b. a global partnership for development; c. greater access to employment opportunities for the poverty-stricken millions in LICs; d. cancellation of bilateral debt and increases in ODA by donor nations to 0.7% of GNP as proposed by Pearson; e. affordable access to drugs essential to control serious epidemic diseases.

The Millennium Declaration claims to be a plan for action. While being statistically specific in terms of poverty and hunger alleviation, access to primary education and clean drinking water, it is less definitive in how the stated objectives are to be realised, by whom and by what criteria progress will be monitored. Objectives such as "to ensure global environmental sustainability" defy any rational quantification. What constitutes or is needed to ensure a sustainable environment is highly contentious, there being little evident agreement particularly between those affluent nations and communities that seek ever greater material prosperity and those who would accept a more modest way of life, prudently conservative of, and willing to share more equitably the Planet's critical and essential resources. Most of the worthy objectives are rewordings or re-phrasings of development objectives proposed in countless earlier reports.

Human Development Index ranking: Year 2001

The principles and criteria from which each nation's Human Development Index (HDI) is calculated were outlined in an earlier chapter. HDI assessments rank Norway number one with an HDI of 0.944. Given the complexity of the HDI calculation it is doubtful if there is significant difference among the nations, predominantly affluent OECD nations, that are ranked highest. Within this OECD group life expectancy at birth ranges from 77 to 79 years, GDP/cap @ PPP from $24,200 (UK) to $34,300 (USA). For all LICs average HDI equals 0.561, GDP/cap @ PPP:$2,200, life expectancy 59 years. Table 4 indicates HDR statistics for India, Nigeria and Zimbabwe:

Table 4
HDI ratings for selected nations

	Life expectancy (yrs)	GDP/cap:PPP $USD	HDI
India	63	2,800	0.590
Nigeria	52	900	0.463
Zimbabwe	36	2,300	0.496

Given the resources available to Nigeria and Zimbabwe, their inferior statistics must be attributed to incompetent governance and resource management.

Armament expenditures (2002) and ODA Commitments

As a comparison with ODA commitments, HDR 2003 presents estimated trade in armaments and the numbers active in armed forces. The data cast a melancholy reflection on international priorities: the willingness to invest extravagantly on armaments, the shameful indisposition among the wealthy to invest to alleviate human suffering.

Table 5
Armament statistics

	Armament exports $ USD billions	Year 2001	Armed forces '000
World	16,500		19,600
U S A	3,900		1,400

Table 6
Overseas aid disbursements 1990 & 2003

	Total ODA 2001 $Bn USD	% GNP 1990	2001	Per cap/year: $ USD 1990	2001
Norway	1.3	1.2	0.83	285	299
U S A	11.4	0.21	0.11	57	39
Canada	1.5	0.44	0.22	83	51
U K	4.6	0.27	0.32	53	80
O E C D	52.0	0.33	0.22	75	63

HDR 2003 calculates that if all OECD governments raised their ODA to 0.7 per cent of GNP, as proposed by Pearson, the total ODA would increase from $52 billion to $165 billion, just 1.0 per cent of world trade in armaments.

HDR 2005

HDR 2005 is subtitled "Aid, trade and security in an unequal world". In his foreword, the then Administrator, Mark Malloch Brown, describes meeting the Millenium Development Goals (MDGs), described in some detailed under *HDR 2003*, as the "single greatest challenge facing the development community". Brown speaks of the obstacles to progress related to aid, fair trade and continued conflict.

The Overview refers to the highly publicised Tsunami catastrophe that left more than 300,000 dead, and the remarkably compassionate and rapid response by private citizens around the world. The willingness to provide assistance to sudden tragedies stands in stark contrast to the unsustained and transitory support for many chronic and lingering sufferings such as the 1,200 children that die every hour from poverty, deprivation and disease. Once the gruesome spectacles inflicted by a natural disaster disappear from the world's television screens, compassionate assistance quickly ends. As Pearson suggested, a weariness with well-doing comes quickly on the heels of compassion for well publicised tragedies. Brown questions whether the many international conferences, discussion and oral commitments to poverty alleviation simply result in rich nations mouthing more pious expressions of sympathy than constructive action.

HDR 2005 proposes three pillars upon which cooperative international action needs to be based:

a.   Improved development assistance – it now suffers from under-financing and is largely uncoordinated and of poor quality (serious inadequacies emphasised by Pearson, Brandt and Brundtland).

b.   International trade under more equitable conditions – During the four years since the WTO Doha 'Development Round', little of substance has been achieved. The rich nations' trade policies persist in denying poor nations a share of international prosperity through fair trade. Inequitable trade policies are inconsistent with MDGs; affluent nations which practise them are unjust and hypocritical.

c.   Security – "Violent conflict blights the lives of hundreds of millions of innocent people" – The UN Secretary-General's report *In Larger Freedom* argues that the lethal interaction of poverty and violent conflict poses grave threats to the collective security of the international community. *HDR 2005* underlines the mutual reinforcement of poverty and deprivation, and conflict.

Of the 52 countries where child mortality is increasing or stagnating, 30 have experienced devastating conflict since 1990. *HDR 2005* condemns donors who deny aid to poor people caught up in nations whose leaders or war-lords are in conflict (the sorry state of the Palestinians, oppressed by their own leaders and kept in a state of subservience by Israel, might have been cited as an example).

In a chapter devoted to aid quality and effectiveness, *HDR 2005* repeats criticism voiced by Pearson, Brandt and Brundtland about the volatility, inconstancy and unpredictability of bilateral aid programmes. The IMF has observed how, over the past 20 years, ODA flows have become more volatile and unpredictable. In consequence poorer nations have no certainty that an aid programme started will be sustained until satisfactory completion. The issue of aid agencies changing their policies and priorities every time there is a change of government or of the agency's senior executive has been discussed.

HDR 2005 reiterates the Pearson, Brandt, Brundtland concern of competition rather than coordination among donors. At last count, Ethiopia was receiving aid from 37 independent donors. Several years ago this author was informed by the minister of agriculture of a small West African nation that during the previous twelve months he had received 289 aid proposals from donors, each offering a different aid programme or project. In varying degrees each donor demands financial and activity reports and evaluations. The administrative load imposed on the recipient's overworked staff frequently costs more in time and effort than the aid is worth.

HDR 2005 emphasizes that violent conflict in one country frequently causes insecurity and conflict in surrounding nations. When the United States and British governments decided to invade Iraq, they were warned that such internationally illegal action would provoke upheaval and instability throughout the Middle East region. These dire predictions have been proved true. Few if any governments in the Middle East are secure from destabilising revolt, while Iraq seems doomed either to civil war or partition into three territories respectively controlled by Kurds, Sunnis and Shias, each hostile to the other two.

HDR 2005 urges the United Nations and its members critically and objectively to evaluate how particular governmental policies and actions have led to exacerbated conflict. It seems unlikely, given the egocentric attitudes of many governments, that any such objective study will ever come about. Those who believe their religious and political ideologies are divinely inspired and the only correct ones for all humanity will never agree to an international study that might question their policies and ideologies.

HDR 2005 sensibly implies there will no end to violent conflict until all governments, ethnicities and religions are prepared to accept the rights of peoples and nations whose beliefs are different than their own; until there is far greater equity in access to all resources and opportunities essential for sustained survival of all humanity.

Final observation

From a critical review and assessment of the World Bank Reports, the UNDP Human Development Reports and many other reputable and reliable sources of information, it is difficult to be even cautiously optimistic that in the foreseeable future the poorest nations and communities will enjoy freedom from poverty, hunger and disease. The melancholy condition, evident among the world's poorest folk, is inevitably aggravated by the alarming spread of HIV/AIDS, by the volatility and unreliability in ODA among affluent nations, a state exacerbated by affluent governments' immense expenditures to sustain the so-called 'War on Terror'. The weariness with well-doing over which Pearson sorrowfully lamented now seems more diverse and pervasive than he could ever have contemplated or countenanced.

4

From Pearson to Johannesburg

Starting in 1967, at ten-yearly intervals, three high-level Commissions, each chaired by a former government leader, were convened by the World Bank. The purpose, as stated by the President of the World Bank was "to study the consequences of [many years] of development assistance, to assess the results, to clarify the errors and propose policies that will work better in the future." The first of the three commissions, led by the Rt. Hon. Lester B. Pearson, former Prime Minister of Canada, began its work in 1967. Subsequent commissions, convened in 1977 and 1987, were led, respectively, by Willy Brandt, former Chancellor of the Federal Republic of Germany, and Dr Gro Harlem Brundtland, former Prime Minister of Norway. Both Pearson and Brandt had earlier been awarded the Nobel Peace Prize.

A difficulty confronted by each commission, indeed by all gatherings that attempt assessments of 'progress in development', is that every agency and person that claims expertise in development expects to be consulted, their concepts and concerns to be entered into the final record. This chapter will offer no more than what appear to be significant and cogent observations and recommendations from each of the commissions.

The Pearson Commission

Supported by a sizeable secretariat the Pearson Commission travelled extensively, held many meetings with governments of LICs, development and donor agencies and with persons experienced in development activities, and reviewed an immense volume of relevant literature. The Commission's report fills some 200 pages with text and tabulations [Pearson 1969]. At the outset it states: "There is far more to development than economic and material progress; [the size] of GNP gives no assurance of possession of other values and qualities".

The report begins with a survey of development programmes from 1950 to 1968. The 1950s and 1960s were a period of growing enthusiasm for assistance to LICs. Between 1961 and 1968 aid flows to LICs increased from $8 billion USD to $12.8 billion. During the 1950s, economic growth in several LICs rose faster than among industrialised nations at a comparable stage of economic development. Toward the end of the 1960s, support for development was flagging and stagnating: "The climate surrounding foreign aid is heavy with disillusion and mistrust." Pearson observed that bilateral assistance was often dispensed primarily to gain political favours, strategic advantages and to promote donors' exports and political ideologies. Aid was given to support and expand the military rather than to alleviate poverty and provide social services. Some donors perceived aid as a means rapidly to stimulate a second industrial revolution among LICs, whereas the greater need was to ensure food, economic and social security. According to Pearson, development assistance should be directed to reduce gross disparities and inequities between the rich and the poor; to prevent the world's peoples becoming more starkly divided between the haves and the have nots.

Pearson stresses that it is illogical to classify all LICs as of a single uniform kind. LICs vary enormously in size, population density, ecologies, needs, opportunities and resources. In 1968, India's population was 530 million (in 2003 more than one billion), the Gambia was home to 300,000 (2003 estimate: 1.1 million). Between 1950 and 1985, 60 former colonies attained political independence. Many, stressed by poverty, had little experience in self-governance, public service administration, financial, resource and industrial management. The world had no earlier experience of political and economic change on so large, rapid and diverse a scale. Assistance programmes were a novelty and largely conducted by trial and error, many donors seeking unrealisable rapid results and simple remedies for highly complex and disparate difficulties.

Pearson recommended, through sensitive consultation between donors and recipients, development objectives be more precisely defined with criteria conducive to meaningful measurement and evaluation of progress or retrogress. (In this author's 40 years of experience, a common weakness among development programmes has been that too little time and effort is devoted to systematic determination of need, opportunities, resources and potential constraints, before committing funds and embarking on development programmes and projects.) All development should be destined to improve the quality of life, health and longevity, with more equitable distribution of wealth and resources. Development and economic growth must become self-sustaining so that LICs eventually become independent of foreign assistance, with objectives more likely to be realised through a patient, gradual unfolding than by seeking quick solutions. When Pearson deprecated sub-Saharan Africa's extreme reliance on foreign aid, often invested in projects with objectives dictated by the donors, he and his Commission did not anticipate that Africa's plight would further deteriorate to its present woeful condition.

Near the beginning, Pearson asks rhetorically: "Why should rich countries seek to help those that are poor?" Pearson's simplest answer for civilised nations is that it is the moral responsibility of those who have in abundance to help those who have not. The fullest, most efficient utilisation of the Planet's resources will result from international cooperation not conflict. Environmental pollution caused by unclean habits in one nation can adversely affect the health and ecological well being of other nations. Disease epidemics do not respect national boundaries. In a related context Pearson notes that, because of sustained international programmes, mortality and morbidity attributable to bubonic and related plagues, cholera, malaria and smallpox had decreased dramatically between 1951 and 1966.

The report repeatedly insists that concepts of instant or rapid development are illusory and lead to disappointment and frustration. With thoughtful foresight, systematic planning, patience and persistence, many nations classed as LICs (in 1968) can achieve self-sustaining growth by the end of the twentieth century. (In the case of several Asian nations, Pearson's prophecy was indeed fulfilled.) The Commission states: "There is an urgent need for a clearer purpose and greater coherence in all development programmes." Technical assistance too often fails to adapt its objectives to the determined needs of recipients. "ODA should be raised to 0.7% of donor GNP by 1975". (As is illustrated in this publication's tabulated data this goal has been realised by very few donor nations.)

Other observations include: "Wealth does not entitle a rich and powerful nation to dominate another country's values and lifestyle as a condition of the aid it has given." Aid to a poorer nation will not guarantee, nor should it, that the recipient will adopt the donor's political doctrines or ideologies. "Unemployment and underemployment have reached critical proportions throughout much of Asia and Africa and without appropriate remedial action will grow progressively worse." "Foreign indebtedness has risen rapidly … debt service increased at 17 per cent a year during the 1960s … Debt relief should be recognised as a legitimate form of aid". "In order to protect their own agricultural and commercial interests, trade policies of advanced nations have raised unfair obstacles to export earnings for many poorer nations."

The report refers to remarkable increases in food grain production brought about by farmers' adoption of higher yielding crop phenotypes of wheat and rice, cultivated under more productive farming systems. In India, annual increase in cereal grain production rose from 0.1 per cent between 1960 and 1965, to 7.2 per cent from 1966 to 1968. Despite many disappointments and imperfections, the record of development assistance is encouraging. During the 1960s, while foreign aid contributed barely 2 per cent to the total income of LICs, where sensibly and productively invested aid was immensely beneficial, leading to increased output from agriculture and other production systems. When well directed, aid has stimulated increased private foreign investment. Aid investments in education and training programmes have served to expand and improve skills and competence in the work force. The drive towards modernisation among LICs has inevitably created conflicts between guardians of long-standing traditions and people who sought adoption of more advanced technologies and social systems. Underlying all development considerations is the need for more equitable sharing of the benefits of progress among individuals and nations. "Sustainable, stable development demands equitable distribution of wealth and access to essential resources."

Rapid population growth imposes serious constraints to stable development. As populations grow, there must be more investment in facilities for education, health services, housing, potable water, hygienic sanitation, and other essential services. During the 1960s, across Asia, Africa and Latin America, urban populations were expanding at double the rate of overall population increase. Pearson pointed to inadequate attention and planning among development agencies and LIC governments, with consequent neglect of the resources and services essential to the needs of large urban communities. Past experience with urban slums during and following the industrial revolution was largely ignored by nations where the slums proliferated and in some instances still exist. Some of the smallest, poorest LICs have been pressed to depend on technologies designed and developed for more advanced countries, technologies ill suited to the recipients' needs and resources. Every nation needs access to a cadre of scientists and technologists competent to choose and adapt technologies appropriate to need. Where governments are willing to cooperate, regional institutes of technology can serve several nations.

The Pearson report presents a long list of recommendations, many still relevant and valid but ignored and unfulfilled: on trade policies, private foreign investment, better systematic planning and criteria of assessment for development programmes and projects, the urgent need for debt relief and low interest loans, no aid in the form of armaments, and total abandonment of tied aid. Tied aid, in Pearson's opinion, is a serious impediment to progressive development: it demands that LICs import goods and services, often of low priority, frequently at a higher cost than if purchased from sources other than the donor. Tied aid most often brings greater benefit to the donor's industries and inhabitants than to poor folk in LICs. Donors should allow recipients to use aid funds to purchase essential materials from wherever is most economic. The report makes 20 recommendations on how to plan and disburse development assistance more rationally and effectively. It urges more investment in education, health and social services, family planning and population control. It proposes reconsideration of various World Bank, IMF and IDA procedures, particularly the need to expand and improve effectiveness of the IDA, the Bank's soft loan window.

The Brandt Commission

The report begins by expressing the hope that the Commission will: "contribute to the development of worldwide moral values" [Brandt 1980]. It speaks of the immense risks that threaten future generations in absence of more determined commitments to international cooperation. Given Brandt's experience during World War II, it is not surprising that his priorities for change and reform are "Peace, Justice, Jobs", in that order. The report "… raises the traditional questions of peace and war … and how to overcome world hunger, mass misery and alarming disparities in living conditions between rich and poor". Brandt repeats Pearson's warning that "quick solutions [to development difficulties] are an illusion." He stresses the need "to curb the mounting spiral of sophisticated and expensive weaponry". When the report was compiled [1980], total ODA accounted for less than 5 per cent of the annual global military expenditure of about $450 billion USD. The cost of a 1978 tank would provide storage for over 100,000 tonnes of grain and thereby reduce substantial post-harvest loss; the price of one jet fighter plane would finance 40,000 rural village pharmacies. Any proposal for a "new economic order" must ensure significant worldwide disarmament.

Despite the technical and financial resources available to increase agricultural production and food availability, chronic hunger persists as a disgrace to all civilised people and its alleviation must be a high priority for all development agencies. "Health care, social development and economic progress must advance interdependently". In other words, 'development' cannot be conceived as a solitary isolated issue but must comprehend interdependent, interactive components in highly complex systems. Truly cooperative consultation between donors and LIC recipients is critical to reliable definition and assessment of development programmes.

Brandt contends that politicians and bureaucrats who determine donor development policies are "too cut off from the experience of ordinary people". Industrial nations tend to behave as if all fuel energy resources are infinitely renewable. There is urgent need for investment to develop alternative renewable energy resources. Atmospheric and other environmental pollutions, excessive exploitation of terrestrial and aquatic resources are spreading across the planet. Over-fishing, uncontrolled deforestation, soil erosion and desert spread threaten the lives and well being of many communities. Affluent nations fail to recognise that we live in a world of common interdependence. Even as their economies have expanded and prospered, donors have made little effort to reach the Pearson target of assigning 0.7 per cent of GNP to ODA. Brandt regrets that the United States, a leader during the early 1960s, had substantially reduced its ODA budget by 1978. "The world's most powerful and wealthy nation should not be content to play a marginal role as other affluent nations seek to promote a more equitable world."

Though Brandt did not try to define 'development', he insisted it should focus on the wellbeing of people and not simply on technologies.

Nor should 'development' be restricted to economic and industrial growth. Wealth and resources generated must be more equitably distributed. Development's purpose is to enable poor nations to realise self-fulfillment, the specific nature being dependent on a nation's history, cultural heritage, religious tradition, resources, climate and ecologies. To reduce widespread unemployment in heavily populated LICs, technologies must be labour-intensive, not highly mechanised. Brandt re-echoes Pearson's concerns for the serious tensions that arise as nations pursue modern technologies.

Brandt's recommended priorities include:
- alleviation of hunger, malnutrition and extreme poverty, particularly across Africa and Asia;
- improvement of conditions for, and removal of constraints to trade in agricultural and other commodities produced in LICs;
- reform of the international monetary system to stabilise exchange rates, with greater provision of Special Drawing Rights for poorer nations; World Bank and regional banks should sizably increase their lending capacities;
- donors must commit their ODA and development assistance on a predictable, long-term basis; as repeatedly emphasised, all development proceeds slowly, aid projects of short duration often do more harm than good;
- an international energy strategy that encourages rigorous conservation and cooperative development of alternative renewable resources;
- a systematic programme for long-term international food security.

The Brundtland Commission

The Commission's name – The World Commission on Environment and Development – defines its purpose [Brundtland 1987a]. Most often quoted is Brundtland's concept of 'Sustainable Development': "To ensure that it [development] meets the needs of the present without compromising the ability of future generations to meet their own needs".

It is a statement to be accepted as a general principle since no one on the Commission or elsewhere could reliably predict what future generations will need, want or demand. Will they aspire to the standards of today's most affluent and extravagant societies, or will they accept a lifestyle more conservative and protective of the Planet's resources and ecosystems?

The chair, Dr Gro Harlem Brundtland, appointed three advisory panels on Energy, Industry and Food security. This author served on the Food Security Panel that composed a complementary report: "Food 2000" [Brundtland 1987b]. The observations and recommendations of "Food 2000" are discussed together with other issues related to agriculture and agribusiness.

Though both Pearson and Brandt referred to harmful degradation of environments, Brundtland was specifically instructed to examine the relation between environment and development. The report states: "The environment does not exist as a sphere separate from human actions, ambitions and needs ... the word environment [has] a connotation of naivety in some political circles". "The word 'development' has been narrowed to the perception of what poor nations should do to become richer". "The 'environment' is where we live; 'development' is what we do in attempting to improve our lot within our abode". "The downward spiral of poverty and environmental degradation incur a waste of opportunities and resources". "Economic growth must be based on policies that sustain and expand the environmental resource base".

Brundtland cites a litany of environmental disasters and degradations:

- Each year six million hectares of productive dry land turns into worthless desert.
- More than 11 million hectares of forest are destroyed every year.
- Acid precipitation (from industrial and vehicle emissions) kills forests, lakes, irreparably damaging vast tracts of land.
- Fossil fuel combustion generates atmospheric CO_2 to stimulate global warming.

- About 60 million people, mostly children, die each year of diarrhoeal diseases caused by infected drinking water.
- Environmental degradation undermines economic development.
- Deforestation by highland farmers causes flooding on lowland farms.
- Deforestation of hillsides damages lowland habitations by soils, no longer bound by tree roots, being released and washed down by heavy rains.
- During the 1970s, 35 million Africans were severely afflicted by drought stress.

Brundtland disparages investments in armaments. Military expenditures (during the 1980s) amounted to one trillion US dollars per year and showed no sign of being reversed. Spending tax revenues on armaments pre-empts resources that could be invested more productively to protect the environment and alleviate poverty. A precise concept of 'sustainable development' is limited by an inadequate state of knowledge, technology and social organisations. No one can predict the long-term consequences of diverse degradations of ecologies and ecosystems, of excessive exploitation of resources that are non-renewable or slow to be regenerated. Science cannot determine the capacity of the biosphere to tolerate the countless adverse effects of human and industrial activities. Sustainable development, globally and internationally, demands that all people and communities, particularly the most affluent, adopt standards of living commensurate with the Planet's ecological and natural resources; in particular to be more conservative in their consumption of non-renewable sources of energy. *Per caput*, affluent nations consume 80 times the energy resources used by sub-Saharan Africans. As LIC demands for energy rise, it is essential that affluent highly industrialised nations use energy more conservatively. Anticipated industrial growth among LICs emerging from reliance on primitive agriculture portends more pollution of the atmosphere, of ground and surface waters if industrial technologies common among the affluent are extensively adopted. Energy derived from combustion of coal, the most abundant fossil fuel, produces excessive emissions of undesirable pollutants. Technologies more conservative of energy and less polluting are critical to the protection of vital ecologies and eco-systems.

Brundtland urges measures to discourage high rates of population expansion among poorer nations. Persuasive evidence indicates that, as people rise out of extreme poverty and become better educated, they choose to limit the size of their families. Biodiversity is seriously under stress with countless species rapidly becoming extinct. Wild species provide drugs, genetic materials for plant and livestock breeders, raw materials for industries. In addition to economic practicalities, Brundtland cites ethical, aesthetic and scientific reasons for biodiversity conservation.

The United Nations Conference on the Law of the Sea drafted an international agreement for conservative management of the Planet's oceans. The Brundtland report states: "Fisheries agreements should be strengthened to prevent current exploitation." A few years later, in the face of widespread international opposition, Norway resumed whaling, the capture and slaughter of seriously endangered aquatic animals. When Norwegian whaling resumed, Dr Brundtland was again Norway's prime minister!

The Brundtland report recommends: "Governments and international agencies should assess the cost-effectiveness, in terms of security, of money spent on armaments compared with investing to reduce poverty, to protect and repair a ravaged environment." Under the heading: "A threatened future", the report states: "The Earth is one but the World is not. We all depend on one biosphere to sustain our lives. Yet each community, each country strives for survival and prosperity with little regard for others. Some consume the Earth's resources at a rate that would leave too little for future generations. The majority live with the pernicious prospect of poverty, hunger, squalor, disease and early death."

In essence, the last paragraph summarises the Brundtland philosophy and proposals. Discussed in greater detail are specific issues including:

a. control of the international economy; b. population and human resources; c. food security: sustaining the Earth's resources and potential; d. species and ecosystems: conservation of resources for development; e. energy: critical choices for the environment and development; f. industry: how to produce more goods and services with less demand on scarce resources; g. the urban challenge: planning and providing for expanding cities and towns; h. peace, security, and a severe restriction on armaments.

Pearson, Brandt and Brundtland

The three Commissions engaged the services of highly distinguished people with extensive relevant experience. The secretariats, consultants and advisers contributed an impressive volume of apposite data and information. Though emphases differ among them, the commissions' observations and recommendations are consistently in agreement. All lay emphasis on the widening gulf between the rich and the poor, both among and within nations. At the time of Pearson [1967], LICs represented 66 per cent of world population with access to barely 12 per cent of total world GNP. Twenty years later, Brundtland reported developing countries' populations as representing 70 per cent of the world total while owning only 10 per cent of global GNP. Brundtland supported the Pearson and Brandt recommendation that donor nations contribute at least 0.7 per cent of their GNP as ODA. Between 1965 and 2000 total ODA as a percentage of donors' total GNP fell from 0.48 to 0.29 per cent. The United States administered an even more dramatic drop: from 0.49 to 0.1 per cent.

All three Commissions were disturbed by rising LICs debt, which rose from an estimated $50 billion USD in 1967 [Pearson 1969], to $250 billion in 1986 [Brundtland 1987]. Total ODA from all donors was roughly $6.5 billion USD in 1967, which rose to $34 billion in 1986. World Bank data indicate that both Pearson and Brundtland underestimated LICs indebtedness. The Bank World Development Report for 1985 states that between 1970 and 1984 LICs medium and long-term debt increased from $70 billion to $700 billion. Among the poorest LICs, debt rose from 19 to 58 per cent of GNP.

All three Commissions spoke of the urgent need for poverty alleviation and despaired of how little related progress has been made. Most of the world's poor lived in south Asia and sub-Saharan Africa. The World Bank estimates that between 1985 and 2000, people who survive on less than $2.00 per day in south Asia rose from 520 million to 1.23 billion, in sub-Saharan Africa from 180 to 570 million, data that indicate one-half the world's poorest folk are south Asians, one-third sub-Saharan Africans. The three Commissions deplored that the richest nations invested so much more in armaments than in assistance to the poor. Brandt estimated that total donor ODA in the 1970s was less than 5 per cent of their expenditures on armaments. The Commissions were in concert in urging more support for health and social services, hygienic sanitation, adequate clean drinking water. The Commissions pointed to the high incidence of unemployment and emphasised the need to create more labour-intensive jobs particularly among poor rural people. Regrettably, both politicians and workers' trade unions among affluent nations energetically voice their opposition to support for employment in LICs, especially where the LIC workers can produce cheaper goods for export to wealthier nations. So much for the plea for workers of the world to unite! The Commissions' demand for fairer international trade particularly in agricultural commodities, with sizable reductions in export subsidies and less visible barriers to trade by Americans, Europeans and Japanese, fell on deaf ears. Brundtland estimated agricultural subsidies by OECD members at *ca* $250 billion per year. Despite pleas from LICs, the World Trade Organisation and others of compassionate disposition, it is estimated that agricultural subsidies increased to more than $350 billion by the turn of the century. Up to the time of Brundtland, at least 50 per cent of LICs' GDP was derived from agriculture, hence the collective call for more support to agriculture and rural agribusiness. The dramatic difference in food grain production between Asia, where harvests had substantially increased from the late 1960s, and sub-Saharan Africa where food production *per caput* had stagnated or declined was cause for alarm.

Because of its declared purpose, illustrated by its title, the Brundtland Commission gave special emphasis to environmental issues. Nonetheless, all three Commissions expressed concern over degradation of arable land, widespread loss of soil fertility, atmospheric, ground and surface water pollution, over-fishing of coastal waters and deep oceans, impoverishment of biodiversity, deforestation, desert spread, excessive consumption of fossil fuels with subsequent rise in 'greenhouse gases' and global temperatures. They reminded those to whom the reports were directed of the interdependence of economic and technological developments with environments and essential resources.

The three Commissions reported on the persistent expansion of cities and urban communities. As is the case among many industrialised economies, urban growth is poorly planned and controlled, too little critical attention being devoted to urban planning, to provision of the services essential to healthy urban communities, to inevitable traffic congestion and resultant air pollution from exhaust emissions.

While illustrating impressive instances of productive growth among several Asian nations, there was a clear consensus that, as mentioned early in the Pearson report, the gap between the rich and the poor was widening. As Pearson remarked in a public address, among the rich nations there seemed to be a weariness with well-doing. All the Commissions were confident that the deprivations described could be remedied if the political will existed. Regrettably the deficient political will apparent to Pearson, Brandt and Brundltland is even more depressingly evident as this text is being composed. The world's most powerful nations display greater dedication to imposing their will on weaker peoples by force of arms than in alleviating poverty and the many gross inequalities evident for so long.

In light of the Commissions' comprehensive catalogue of poverty, hunger, disease, of other human deprivations and suffering, one wonders why so many costly international conferences and gatherings were subsequently convened simply to reiterate and re-emphasise the Commissions' observations and recommendations. It would be interesting to learn how many of those who voiced their pious platitudes at subsequent conferences, indeed how many persons presently employed by development agencies, have read the reports of the three Commissions. An account of two such extravagant conferences – the first in 1992 in Rio de Janeiro, the second in 2002 in Johannesburg – constitutes the remainder of this chapter.

International conferences on environment and development

'Environment' is derived from an Early English word meaning 'around' or 'to surround'. Later, the verb 'to environ' was synonymous with 'to encircle', while the noun 'environ' designated the surrounding districts of a town. The noun 'environment' was coined in the early seventeenth century to mean 'that which surrounds or encircles', sometime later: 'the conditions under which an organism lives or something is developed'. In contemporary vocabularies it is used to describe all that circumscribes or influences an activity, a discipline, an ecology, or a state of existence. People speak and write of political, economic, social, cultural, technological, industrial and physical environments. 'Environment' may now be interpreted as everything that impinges upon, conditions or influences a state of existence or a development activity. In a more limited sense it relates to physical environments: the Planet's surrounding atmosphere, structural components, resources, climate and weather, and how these influence and are influenced by natural phenomena, humans, other organisms and their activities.

The first United Nations environment conference was convened in 1972 in Stockholm and led to the establishment of the United Nations Environment Programme [UNEP] with headquarters in Nairobi. UNEP's stated responsibilities are to 'encourage sound environmental practices, to study and monitor international environmental conditions around the world, and to recommend alternative energy sources'.

As 'environment' is so vast and all embracing, during its early years UNEP had difficulty in defining and pursuing unique objectives not covered by other international agencies. A second UN environment conference, held in 1987 in Nairobi, did not add significantly to existing knowledge or concepts.

United Nations Conference on Environment and Development (UNCED)

UNCED, also designated "the Earth Summit" and convened in 1992 in Rio de Janeiro, was the largest such conference ever held. More than 100 national political leaders, each with a sizeable delegation, participated. Also present were several hundred representatives from UN bodies, other international organisations and associations, academic and scientific bodies, together with persons interested in one or more of the phenomena, activities and interactions related to the earth's environment and its diverse components. A parallel conference, attended largely by interested NGOs, was convened in Rio concurrently with UNCED.

This author's attempts to discover the cost of this enormous event were unsuccessful. Considering that many large delegations were led by heads of governments, the total cost for travel and accommodation would run into more than a million dollars. Added to this would be the cost of conference facilities, secretariat's time and resources; time, effort and resources invested in convening and reporting on the many conferences, meetings and consultations, and the enormous volume of documents that preceded and prepared for UNCED. A cost versus benefit analysis would defy the ingenuity of the most experienced accountants.

Several conferences and meetings were convened ahead of the UNCED.

The ASCEND Conference

An Agenda of Science for Environment and Development into the twenty-first century (ASCEND), convened by the International Council of Scientific Unions (ICSU) during November 1991 in Vienna, was among the most impressive of pre-UNCED gatherings. ASCEND was attended by some 240 participants and 37 observers from 70 nations. The published proceedings, which run to roughly 250,000 words, include 10 keynote speeches with subsequent presentations on: Population and Natural Resource Use; Agricultural Land Use and Degradation; Industry and Waste; Energy; Health; Hydrologic and Biogeochemical Global Cycles; Atmosphere and Climate; Marine and Coastal Systems; Fresh Water Resources; Biodiversity; Quality of Life; Science and the Environment; Policies for Technology [ICSU-ASCEND 1992].

As a further contribution, the ICSU Commission on the Application of Science to Agriculture, Forestry and Aquaculture (CASAFA) was commissioned to compose a report on Sustainable Agriculture and Food Security, a document of some 56,000 words [ICSU-CASAFA 1991]. The ICSU documents were two among an immense volume of papers tabled in Rio. In light of the enormous volume of literature, it is doubtful if any delegation found time to read all of the UNCED conference documents. One suspects that many were read in their entirely by few other than original authors.

The ASCEND conference was convened to define a twenty-first century agenda for the natural, social, engineering and health sciences as they relate to environment and development. The ASCEND recommendations provide a useful relevant perspective for the future direction of international science. Under the challenging heading "What can science do?" the ASCEND report points out that Earth is the only celestial body where human life is known to exist and will remain habitable only as long as the Planet maintains its unique life-support resources. The land, oceans and atmosphere are interconnected and interact through biogeochemical cycles of water and nutrients that together form the earth's highly complex state. Investigations have shown how human activities have changed the planet's life-support system and put sustainable development of present and future generations at serious risk.

The principal adverse human activities and interventions include: rapid population growth, excessive resource exploitation, agricultural and industrial technologies, fossil fuel combustions that disturb the planet's climatic systems. ASCEND urged greater support for scientific research on the planet's complex systems; research on natural fluctuations and those induced by anthropogenic forcing, on hydrological cycles and their interactions with soils and vegetation, the effects of climate change on natural and managed ecosystems, on coastal zones where 75 per cent of the world's inhabitants live, on means to assess loss of biodiversity, protection of fragile ecosystems, assessment of the earth's carrying capacity, determination and design of systems for environmental forecasting, changing land use, waste and pollution related to public health, discovery of new, more efficient energy sources.

Recommendations from the Earth Summit: Agenda 21

The extent, complexity and diversity of the UNCED proceedings defy simple, concise summary. Among a mass of proposals, recommendations and tentative agreements, UNCED received general consensus among government delegations for International Conventions on (a) Climate Change; (b) Biological Diversity; (c) Control of desertification in vulnerable ecologies. UNCED proclaimed 27 Principles, many addressed to national governments. Principle No 1 states that human beings, the centre of concerns for sustainable development, are entitled to a healthy and productive life in harmony with nature. Other principles call for: development to be fulfilled so as to meet the needs of present and future generations (an echo of Brundtland); all States to cooperate in eradicating poverty; all nations to cooperate in protecting and restoring the integrity of the earth's ecosystems; international cooperation in exchange of scientific and technical knowledge; all citizens of the world to be adequately informed and encouraged to protect the planet's environment and resources; governments to enact environmental legislation relevant to their own conditions; governments to enact legislation to determine liability and provision of compensation for victims of pollution and damage to property (More than 4,000 years ago, Hammurabi, the King of Babylon, decreed that, if irrigation water or animals escape from one farm and cause damage to another farmer's property or crops, the guilty farmer shall make adequate compensation to the victim)[Harper 1999].

Other Agenda 21 Principles call for: assessment of activities that adversely affect the environment, that governments provide timely notification of events or activities that may cause injury to the property or resources of their neighbours. Agenda 21 emphasises the vital role of women in resource and environmental management. Finally, it enunciates principles woefully ignored by several nations with powerful military resources. Principle 23 states: "Environment and resources of people under oppression, domination and occupation shall be protected." A principle with particular relevence to such oppressed people as the Palestinians and Iraqis.

Principle 24 states that warfare is inherently destructive of sustainable development and that all governments should have respect for international law and protect the environment in times of armed conflict. Principle 25 succinctly states that: "Peace, development and environmental protection are interdependent and indivisible."

The 1992 Earth Summit was much larger and more ambitious than earlier conferences related to environment. Its major theme was described as 'Sustainable economic development', which may be interpreted as development that does not excessively use or destroy natural resources at levels unsustainable for the foreseeable future. Agenda 21, the overall plan, called for large countries to develop industries and technologies that are protective of the environment and its resources.

A Special Commission was appointed to ensure that countries were faithful to the commitments and promises reflected in Agenda 21. In common with all UN agencies and similar international bodies, the Commission wields no power to enforce nations to live up to their promises. It was hoped that monitoring and dissemination of relevant progress made after Rio would encourage all nations to respect and adhere to the Rio Principles as set out in Agenda 21. At its first meeting in 1994, the Commission reported that several industrialised countries were reneging on their Rio commitments, and were providing less than one-half the funds promised. Governments of the most prolific atmospheric polluters appear more interested in protecting the profits of their polluting industries, than in investing in technologies that would reduce emissions from smoke-stack industries and automobiles.

The recommendations in Agenda 21 in large part re-emphasise those of earlier commissions and commentators. It was stated that Agenda 21 is intended as a blueprint for action to ensure environmentally conservative and sustainable development throughout the twenty-first century. UNCED offered several policy recommendations to the poorer nations:

a. to expand processing, distribution and marketing capabilities (i) to satisfy changing internal needs; (ii) to be competitive in international trade

b. to pursue diversification to reduce dependence on a limited variety of commodities;

c. to adapt technologies that militate against excessive exploitation of natural resources.

It goes on to state that sustainable development demands prudent investment from both domestic and external financial sources. Negative net transfers of financial resources (i.e., external payments for imports and debt servicing being in excess of income from international trade) exacerbate poverty, resource depletion and environmental degradation. Servicing of external debt imposes an impediment to economic and social development. Externally-imposed structural adjustment programmes, while sometimes desirable to improve balance of payments accounts and fiscal budgets, often cause adverse social and environmental conditions: restriction of allocations to health care, education and environmental protection.

It is difficult to determine to what extent the Earth Summit stimulated public awareness of, and support for, actions designed to protect the planet's environment and its resources. Though UNCED amassed an enormous volume of information and opinion, one might conclude that, for the most part, Agenda 21 repeated and enlarged upon observations and recommendations published in the Pearson, Brandt and Brundtland reports. One might speculate that the enormous cost incurred in organising and convening UNCED could have been more productively invested in programmes dedicated to poverty alleviation, environmental protection, biodiversity and resource conservation. In the early 1960s, the national committee of the Canadian Freedom From Hunger Campaign classified its volunteers into two categories, 'the talkers' and 'the doers'. One might question if UNCED and other huge international conferences mainly provide a forum for 'talkers' rather than opportunities for 'doers'. One person who participated in UNCED described it as a dialogue of the deaf: typically, the political leaders came to proclaim their resounding rhetoric but not to listen to anyone else's point of view.

From Rio to Johannesburg

In a thoughtful publication M.S. Swaminathan reviews the Rio recommendations, what has subsequently resulted, and his hopes for the Johannesburg conference held in 2002 [Swaminathan 2002]. In his introduction he states: "The environmental policies advocated by the richer nations are designed to protect their high standards resulting from the unprecedented growth in exploitation of natural resources during the last century ... the poor nations, in contrast, are faced with the need to produce more food from hungry and thirsty soils ...". Swaminathan re-emphasises Pearson, Brandt and Brundtland's plea for effective actions to provide stable paid employment for the millions of unemployed in the LICs.

Swaminathan refers to several international gatherings which have supported and amplified some of the Agenda 21 Principles and recommendations: a. the 1995 FAO World Food Summit, whose objective was, by the year 2015, to reduce by half the number of chronically malnourished people, estimated at over 800 million, 28 per cent of whom are sub-Saharan Africans; b. the 1997 Kyoto meeting of national governments that proposed by year 2010 a 15 per cent reduction in 'greenhouse gases', an objective that would reduce CO_2 emissions by 800 million tonnes. It is distressing but not surprising that, while EU nations are generally supportive, the United States has not formally signed on to the Kyoto Convention.

Swaminathan [2002] discusses, and illustrates with a wealth of statistics, environment and development related to specific issues including: food and water security; environment and the citizen; integrated gene management; ending food insecurity and poverty; nutrition in the third millennium. He concludes by recommending the actions needed, mainly as these relate to the alleviation of hunger, well illustrated by India's experience and progress, in particular his own exceptional contributions during the past 50 years, inspired by the compassionate concept: "the test of our progress is not whether we add more to the abundance of those who have much; it is whether we provide enough for those who have too little."

The UN World Summit on Sustainable Development [WSSD]

The WSSD was convened in 2002 in Johannesburg ostensibly to review and determine progress since 1992 in effective implementation to the Rio Earth Summit Principles and recommendations set out in Agenda 21. The published proceedings of WSSD consist of two documents: the Political Declaration; the Plan of Implementation.

The Political declaration

This document begins with the reaffirmation by the assembled nations of their commitment to sustainable development, to the building of a humane, equitable and caring international society, cognisant of the need for human dignity for all the world's peoples. It assumes a collective responsibility to strengthen the 'interdependent pillars of sustainable development' – economic, social and environmental protection.

The document does not suggest how, quantitatively or qualitatively, effective responses to Agenda 21 proposals will be assessed. Statements under "The challenge we face" suggest that little progress has been made towards eradication or significant alleviation of inequities between rich and poor, or of the environmental degradations described by Pearson, Brandt and Brundtland and reiterated at the Earth Summit: (a) "The deep fault-line that divides human society between the rich and the poor, and the ever increasing gap between the developed and the developing world pose a major threat to global prosperity, security and stability." (b) "The global environment continues to suffer; loss of biodiversity continues, fish stocks are being depleted, desertification claims more and more fertile land, the adverse effects of climate change are already evident, natural disasters are more frequent and more devastating, developing countries are more vulnerable, air, water and marine pollution robs millions of a decent life." It further states: "We risk the entrenchment of these global disparities; unless we act in a manner that fundamentally changes their lives, the poor of the world may lose confidence in their representatives and the democratic systems to which we remain committed."

The Political Declaration restates the threats to sustainable development of the world's people – chronic poverty, hunger and malnutrition; foreign occupation; armed conflicts; corruption; natural disasters; intolerance and incitement to racial, ethnic, religious and other hatred; chronic diseases such as HIV/AIDS, malaria and tuberculosis. It urges that developed countries commit agreed levels of Official Development Assistance (as proposed by Pearson in 1969).

From these comments, and though not specifically stated, one can only conclude that the millions of dollars invested to convene and conduct the Earth Summit have resulted in few if any of the results hoped for or expected. The weary litany of persistent ills, inequities and degradations described in reports from UNCED, other international conferences and commissions, and the actions needed for their eradication or alleviation have been substantially ignored by the world's most affluent and powerful nations.

The Plan of Implementation

This dense document of some 35,000 words contains 153 recommendations, several with additional supporting paragraphs. For example, *Poverty Eradication*, which again repeats that "eradicating poverty is the greatest global challenge facing the world today", includes targets and objectives set out in Agenda 21, recommendations from other international conferences and the UN Millennium Declaration, all laid out in six main paragraphs with 38 sub-paragraphs. Other proposals for implementation relate to "changing unsustainable patterns of consumption and production"; "protecting and managing the natural environment", which includes over 160 recommendations covering a vast diversity of objectives.

The Plan requests substantial financial and political support for many existing international organisations, conventions and programmes, and for a variety of proposed new programmes: "sustainable development in a globalising world"; "health and sustainable development"; "sustainable development of small island states"; "sustainable development for Africa", which makes specific reference to the New Partnership for Africa's Development (NEPAD), which in turn repeats many required actions stated in earlier documents. The Plan cites other desirable sustainable development initiatives for Latin America, the Caribbean, Asia and the Pacific, West Asia and the region covered by the Economic Commission for Europe.

The Plan of Implementation ends with a "Means of Implementation", roughly 130 recommended decisions and actions demanded of the world's nations. The ambitious magnitude and all-embracing diversity of this document would strain the credibility of the most devoted supporter. It repeats demands for nations to increase their levels of ODA, demands that have been consistently ignored for more than thirty years. Most recommendations are presented in broad unspecific terminology with no attempt to define words amenable to widely different interpretations, such words as 'development', 'sustainable', 'globalisation', 'good governance', 'transparent', several of which Pearson, Brandt and Brundtland found difficult to define. Contrary to traditional wisdom and management doctrines, that a Plan must be presented as a definitive statement of specific priorities, neither the 'Plan' nor the 'Means' present any order of priorities. Most paragraphs are worded in such broad generalisations as to defy clear interpretation for action. Perhaps in the minds of the reporting authors, all proposed recommendations are of equal importance, thus providing donors with an excuse for escape: "We can't possibly support all of these proposals; since there is no order of priority, either we won't support any or only those few that suit our political priorities and ambitions."

The Johannesburg document appears as a record and regurgitation of virtually every recommendation and suggestion proposed by the organisations, agencies and individuals who participated, together with a repetition of issues raised by many conferences and commissions that preceded Johannesburg. The absence of precise definitions, any logical sequence of precise priorities, the human and material resources essential to their effective pursuit, an estimate of the probable time required and cost of each proposed development programme, collectively render the proposals virtually impossible of effective implementation. Even if all donor nations raised their ODA to 0.7 per cent of GNP, the total commitment would fall far short of the cost of effective implementation of all that the Plan proposes.

Final observation

It is difficult to discern what productive practical purposes are served by such mammoth international conferences as those convened in Rio and Johannesburg. The proceedings are replete with resounding rhetoric but woefully lacking in defining out of all that is deemed desirable, what are the most urgent priorities, and by what means should they be actively pursued. In this author's opinion, the enormous investments required to finance these extravaganzas could be more productively devoted to projects designed to alleviate some of the ills and inequities clearly defined by Pearson, Brandt and Brundtland and so often repeated by subsequent conferences and committees.

It is now proposed that yet another international conference be convened to debate and discuss progress in what was recommended at the Rio and Johannesburg. What more lucid insights may be gained from another conference that are not already abundantly evident defies the imagination. Those who are making the proposal might be helpfully reminded of the Harvard economist John Kenneth Galbraith's observation in his classic publication, *The Affluent Society*: "We live in a world increasingly manipulated toward the endless (and mindless) consumption of all we need least."

5

Poverty

Poverty and development

Few persons and agencies dedicated to and experienced in economic and social development would disagree that poverty and its alleviation is a fundamental purpose of all development programmes. Different dictionaries and social commentaries define 'poverty': 'the condition of owning neither wealth nor material possessions'; 'having insufficient income and resources to sustain healthy survival'; 'suffering a lack of basic human needs: adequate food, clothing, housing, clean water, health services and hygienic sanitation'. These quotations define 'absolute' or 'near-absolute' poverty and destitution. Except for a few extreme cases, among affluent nations poverty exists as 'relative' rather than as 'absolute' poverty, where certain segments of society own less and receive lower incomes than the majority. The world's poorest people, most in Africa and Asia, are so poor they struggle to survive.

Defining poverty

To determine and agree upon a precise, internationally recognised definition of 'poverty' is a difficulty that confronts all development, government, non-government and social service agencies. If accepted that those classed as LICs are the poorest nations, in 1982 and 1995 respectively, people with less than $410 and $765 per cap/year would be considered below the poverty line. An enquiry addressed to several international development agencies in early 2004 revealed that, for some the poverty line is set at $1.00/Cap/day ($365/Cap/year) for others $2.00/Cap/day ($730/Cap/year). Several United Nations agencies now regard those whose income is less than $1.00 USD per day as poverty-stricken.

In the United States the 'Official Poverty Line' is an income level determined by the Bureau of Census: poverty classifications among US citizens depend upon in what degree their estimated incomes fall below the Official Poverty Line. In Canada, definitions of what constitutes the poverty line vary among government, non-government and welfare agencies. By some agencies the poverty line is set at the estimated median income for all Canadians. According to the Canadian Council on Social Development, the poverty line will vary according to family size, the location and population dimensions of the community in which the family lives. In the year 2002, the poverty line for a family of four persons could range from $36,250 CAD ($30,000 USD) to $25,000 CAD ($21,250 USD), dependent on whether they lived in a large city, a medium town or a rural community. It is apparent that those classed below the poverty line in Canada suffer far less abject deprivation than Africans and Asians who fall below defined poverty lines of $2.00 or $1.00/Cap/day. The poor of Africa and South Asia, unlike Canadians, do not enjoy access to beneficial health and social services. Millions of poverty-stricken Asians and Africans would be delighted to enjoy the standard of living, social and health services available to many Canadians classed as below the poverty line.

The poverty literature

Canadians were awakened to the existence of widespread abject poverty by two series of broadcast lectures during the 1960s: in 1961 by Barbara

Ward [Ward 1961] and in 1965 by John Kenneth Galbraith [Galbraith 1965]. The dismal situation and the means of alleviation cited by Ward and Galbraith are re-echoed in several of Professor M.S. Swaminathan's perceptive publications [Swaminathan 1999, 1999a, 2002]. The poverty described by Ward and Galbraith 40 years ago remains all too pervasive in 2006.

Galbraith states: "The first and most obvious consequence of poverty is social and civil instability", a truism woefully ignored by contemporary political leaders. Galbraith vividly describes how poverty pursues a downward cycle: poverty begets more poverty since the poor cannot take advantage of remedial opportunities or adapt productive innovations that incur adverse risks. In contrast, wealth begets more wealth and enhances the power of the wealthy to exploit those too poor to protect themselves.

Barbara Ward, in her brilliant historical review of human progress, her Massey Lectures on *The Rich Nations and the Poor Nations,* defines the political, socio-economic, biological and technical actions required to alleviate poverty:

a. Equal opportunity and equitable access to resources among people and nations without distinction of class, race, religion or sex.

b. Universal access to resources and services essential to survival, health and security.

c. Food production and distribution must keep pace with population increase and changes in demographic distribution.

d. Transformation from artisanal crafts to industrial technologies; encouragement and support for rural employment in agribusiness; technological development relevant and appropriate to the needs, resources and opportunities of those who are to apply and benefit from them.

It is often difficult to differentiate between causes and consequences of poverty, in many instances the two being mutually reinforcing, as described by Galbraith. Chronic malnutrition impairs the ability to learn and to work, which in turn deprives those so afflicted of opportunities for stable, gainful employment. Histories record how relative poverty can degenerate into absolute poverty when those with wealth and power oppress and enslave those too weak to resist. Among several ancient civilisations, enlightened rulers, such as Hammurabi, enacted laws to protect the poor from predations by the rich and powerful [Harper 1999].

Extensive discussions of poverty, its attributes, causes and consequences, are published by the International Development department and the Institute of Poverty Research at the University of Manchester in England.

Magnitude and extent of poverty

The UN *Human Development Report* for 2002 estimates that at least 1.2 billion – 20 per cent of total world population – struggle to survive on less than $1.00/day. During the 1990s, if one excludes the Peoples' Republic of China, the numbers in the less than $1.00/day category increased by 28 million. Only 30 of the 155 countries classed as LICs or nations in transition from 'developing' realised *per caput* income growth of more than 3 per cent/yr. The following two tabulations show per cent population with less than $1.00/day (Table 7); relevant regional data during the 1990s (Table 8):

Table 7
Per cent population below $1.00/day

Sub-Saharan Africa	49	Nepal	38
Nigeria	70	Bangladesh	36
Ethiopia	80	India	35

In Europe the first organisations to provide succour to the poor and destitute were the medieval monasteries. In 1597 the English Parliament enacted a Poor Law that elaborated on earlier Elizabethan legislation to assist the indigent, including a compulsory Poor Rate in 1563, and the method by which to collect it in 1573. During the

Table 8
Income growth and poverty reduction 1990s

	Income growth Avge/%/cap/year	Poverty reduction %	Number [M] below $1.00/day		% below $1.00/day	
			1990	1999	1990	1999
S S Africa	—[0.4]	—[1.6]	240	315	47	49
E Asia/ Pacific	+ 6.4	+ 14.9	485	280	31	16
South Asia	+ 3.3	+ 8.4	505	490	45	37

sixteenth and seventeenth centuries town councils in England were required to provide for various forms of social distress: homes and schools for children, hospitals for the sick, relief for the aged, work places for those willing and able to work, correctional institutions for people classed as deliberately idle [Bindoff 1950]. The 1795 Speenhamland Act revised English relief to the poor by delegating to local Parish Councils the responsibility for supplementing rural labourers' wages to a subsistence level.

During the late eighteenth century, about the time of Adam Smith's *Enquiry into the Wealth of Nations* and Jeremy Bentham's *Theory of Legislation*, radical opinions proposed by the Rational Dissenters advocated that Britain's Poor Laws cushioned the poor and incited idleness [Plumb 1963]. Extreme contradictions between those of social conscience who regard assistance to the poor as a government and civic responsibility, and their adversaries who regard poverty and related destitution as self-inflicted, or as divine retribution for sinful malfeasance, are abundantly evident in today's world. Among nations who enact and enforce benign social policies, the gap between the most and the least affluent is narrower than where successful acquisition of extreme wealth is regarded as just reward for diligent endeavour, irrespective of how the disproportionate wealth is acquired. Among the most acquisitive, Francis Bacon's aphorism that 'knowledge is power' has given way to 'wealth is power'.

Poverty and World Bank Development Reports

Since 1980 at the start of each decade the World Development Report has been devoted to Poverty Alleviation. The 1980 Report was entitled *Poverty and Human Development*, the 1990 Report simply *Poverty*, while *WDR 2000* summarises proposed strategies and actions to reduce poverty, indeed how concepts of human development have changed and become more complex over the past 50 years. During the 1950s and 1960s much emphasis was placed on large investments in physical capital and infrastructures. In the 1970s it was realised that physical capital was insufficient and that investments in health, education and training were essential for people to rise out of poverty. The 1980s witnessed a further shift in emphasis consequent upon the growing debt crisis among poorer nations, together with a worldwide economic recession. Evident was a sharp contrast between progress in the East Asian – Pacific nations on the one hand, and the persistent poverty in South Asia and sub-Saharan Africa. *WDR 1990* called for improved economic management, greater reliance on market forces and investment in health and education.

The *World Development Report 2000* estimates that of the planet's 6 billion inhabitants, 2.8 billion (47 per cent) try to survive on less than $2.00 USD/day, 1.2 billion (20 per cent) on less than $1.00/day. Of these poorest, 44 per cent live in South Asia, 24 per cent in sub-Saharan Africa. Gross inequities between the most affluent and most indigent are reported from many nations. *WDR 2000* estimates the percentage of total national income received by the poorest 20 per cent and 10 per cent, and the richest 20 per cent and 10 per cent in each of the nations listed in Table 9.

The 'Factor [Richest:Poorest]' columns show the per cent of total national income received by the richest 20 per cent and 10 per cent divided, respectively, by the per cent of total received by the poorest 20 per cent and 10 per cent. From these data, the 20 per cent richest in Brazil receive 25.5 times the income of the poorest 20 per cent; the 10 per cent richest in South Africa receive 42 times the average income

Table 9

	Poorest		Richest		Factor [Richest:Poorest]	
	20%	10%	20%	10%	20%	10%
USA	5.8	2.3	47.7	32.3	8	14
UK	6.6	2.6	43.0	27.3	6.5	10.5
Canada	7.5	2.8	39.3	23.8	5.2	8.5
Brazil	2.5	0.9	63.8	47.6	25.5	53.0
India	8.1	3.5	46.1	33.5	5.7	9.6
South Africa	2.9	1.1	64.8	45.9	22.3	42.1
Russian Fedn.	4.4	1.7	53.7	38.7	12.2	23.0

of the poorest 10 per cent. It is estimated that the total assets of the world's 350 richest people exceed the combined total GNP of the countries that are home to almost one-half of the world's poorest folk. A recent report (2000) by the UN World Institute for Development Economics states that the richest 1.0 per cent of the world's adults owned almost 40 per cent of global assets.

WDR 2000 outlines the many dimensions of, and difficulties in quantifying poverty, be it individual or family incomes or estimated levels of consumption. GDP *per caput* in $USD or as Purchasing Power Parity, the quotient of national GDP or GNI divided by the size of population provides little indication of the spread of income across any population.

Poverty in India

WDR 2000 discusses the dimensions and causes of poverty with particular reference to India. The Report describes the desperately poor as people a. with income and assets insufficient to acquire such basic necessities as adequate food, shelter, clothing, health services and education; b. who are voiceless and powerless to communicate with or influence those in authority; c. who possess inadequate material and physiological reserves to withstand sudden adverse shocks: disease epidemics, food shortages, extreme climate change. In India poverty, accompanied by severe deprivation of access to education and health services, is in part attributable to inequities among castes and classes and between the sexes. Data from India's National Sample Surveys (NSS) indicate that despite healthy national economic (GDP) growth, of about 6.1 per cent/year during the 1990s, the pace of poverty reduction across India slowed down. While it is extremely difficult to assemble and analyse demographic and economic data from so large and diverse a population, NSS statistics indicate a widening income disparity particularly between a rising urban 'middle class' and urban and rural poor. *WDR 2000* suggests that comparative assessments of urban and rural poverty across India may be distorted by inadequate research methodologies, and that there may be less difference between rates of rural and urban poverty alleviation than some published data indicate. Nevertheless, though the accuracy of available statistics may be questionable, the inequitable distribution of purchasing power between those who enjoy abundant affluence and those who suffer abject poverty is evident to any frequent visitor to India, Africa and other Asian nations.

Marked disparities in the extent of poverty and between rich and poor is apparent among India's states. Over many years, the State of Kerala, which has provided comparatively superior education and health services, enjoys average life expectancies higher than the District of Washington D.C., despite significantly different average income levels.

Alleviation of poverty

WDR 2000 claims to benefit from earlier accumulated evidence and experience and proposes that poverty be attacked in three ways: promotion of employment opportunity; facilitation of empowerment; enhancement of security. The poorest people are deprived of the most critical and essential of opportunities, namely access to stable remunerative employment. *WDR 2000* lists inadequate access to other opportunities that contribute to rural poverty: credit, roads and communications, reliable electrical power, profitable markets, schools, safe drinking water, hygienic sanitation and health services. In the Bank's judgment, poverty is a complex of many unavailable and inaccessible opportunities, and economic status must be assessed by a mix of quantitative and qualitative criteria. To alleviate poverty, governments must act to reduce inequalities in access to essential opportunities and resources. Sadly, alleviation of poverty by provision of more equitable access is slow to come about, as the rich and powerful exert greater influence over politicians and government policies than do poorer citizens.

Empowerment

WDR 2000 states, as an unspecific generalisation, that empowerment requires participation of all citizens: the poor, middle class, the rich and powerful in formulating decisions that will influence national and individual survival and development. 'Empowerment' is a popular word among people who write about social and economic development, a word of no precise meaning. Literally, empowerment means 'giving power to'. Does the Bank's proposal mean that

a. all citizens actively participate in decision-making and intend that there be significant decentralisation of power, authority and responsibility from national governments to local communities?

b. local communities be granted the resources to pursue and fulfill their decisions?

c. on all major issues there be a formal Swiss-style referendum?

d. all people must rely on their nationally elected or appointed governments to enact and enforce legislation that they, the government officials, decide is in the greatest public interest?

In so many self-styled democracies, rich and poor, 'the greatest public interest' is in practice interpreted as what is in the greatest interest of those who enjoy political power and those who directly support them financially and/or ideologically.

Security

WDR 2000 describes 'security' as 'reducing vulnerability to economic shocks, natural disasters, ill-health, disability and violence ... it requires effective national action to manage the risk of economy-wide shocks and effective mechanisms to reduce the risks faced by poor people ... improving the assets of poor people, diversifying activities and providing insurance mechanisms to cope with adverse shocks.' *WDR 2000* further states that among these there is no hierarchy of importance, all of the elements being complementary. "Helping poor people to cope with shocks and managed risks [enables] them to take advantage of emerging market opportunities".

WDR 2000 advocates a comprehensive approach but admits there is no simple all-embracing blueprint for poverty alleviation; each country must prepare its own policies that reflect national priorities, communal and individual needs, resources available and what actions and investments are practically possible. *WDR 2000* encourages private investment, technological innovation, support of micro-enterprises and small business, access to affordable credit, training in technology and business management, creation and maintenance of essential infrastructures. It recommends that poor nations support agriculture, essential services and expand into international markets. *WDR 2000* reiterates and insists that women in all nations and communities be given more equitable rights and opportunities.

Though there is little in the *WDR 2000* recommendations with which to disagree, they cover a vast spectrum and are stated in broad generalisations. Among 'opportunities' listed, access to stable remunerative employment must rank as a high priority. A statement attributed to a British Indian Army officer in 1856 – "Indian famines are famines of work, not of food. Where there is work there is money for food" – was eloquently elaborated by Mahatma Gandhi in the early 1920s when he wrote: "To a nation starving and idle, the only form in which God dare appear is as work and a promise of food as wages".

The Bank authors mention the need for credit but do not emphasise that those who operate and are employed in rural agribusiness, as producers, processors and/or vendors, need available access to technical, financial and marketing advisory services, services that are permanently accessible in their communities not in some distant government laboratory or delivered in the form of brief, infrequent, transitory visits.

The litany of potential insecurities defies interpretative action. Not even the most affluent, scientifically and technologically advanced nations can predict or insure against all possible sources of natural catastrophe: earthquakes, droughts, tempests; of disease epidemics among humans and farm animals; of civil unrest and military conflict. How can equitable rights and opportunities for women be ensured in a male-dominated world where so few women hold high office in governments, in industry or commerce? *WDR 2000* does not suggest how to change communities where religious and/or social traditions deny women their rights to any semblance of equality or equity with their fathers, husbands and brothers. Nor does *WDR 2000* suggest how the rich and powerful in any nation can be persuaded or forced to be sensitively and practically sympathetic to the needs and opinions of their less fortunate neighbours, to legislate for equitable access to resources and opportunities among all citizens.

Neglected opportunities for women

Among many nations, women are demonstrably more prudent and effective managers of money and business affairs than are men, a fact tragically ignored in many development programmes. Until the nineteenth century 'Economics' in English meant management of the household, the same meaning as the Greek word οικονομεια (oikonomeia) from which it is derived. Women, who efficiently managed their households, were the first economists. Throughout most of Africa and much of Asia it is women who manage and control the markets. Some time ago this author proposed to an African finance minister, a man with a PhD. in macro-economics, in order to realise national prosperity, his most constructive action would be to hand over his office to the principal 'market mammy'.

Constraints to trade and hopes for reform

Exacerbating poverty among poorer nations, *WDR 2000* estimates that tariffs and subsidies by OECD governments cause annual losses in welfare of over $20 billion USD, equivalent to 40 per cent of all aid in 1999. Past experience is not a cause for optimism that "the developing world and the international community will work together and combine their experience and knowledge to ensure more rapid progress towards ending poverty during the twenty first century". Discussions reported from a recent meeting of the World Trade Organisation offer hope that nations' subsidies to agriculture, a serious impediment to equitable international trade, may gradually be reduced and eventually eliminated. Recognising the political influences exercised by American and European farmers, it is unrealistic to expect any significant reduction in the $350 billion expended annually to subsidise agricultural exports from the United States and the EU.

WDR 2000 ends by stating that concepts of development cooperation are being reformed, new patterns of aid assistance are being devised and tested. The Bank's stated belief in a revived commitment by the international community to alleviate poverty is hardly borne out by present available evidence. During the 1990s among most OECD nations' ODA as a per cent of GNP was in decline. The world's most powerful and affluent nation invests many billions of dollars annually in armaments and military adventures, less than $12 billion in foreign aid, most of which goes to a nation which by any rational definition cannot be described as low-income, underdeveloped or developing.

According to the International Institute of Strategic Studies, in 2003 total world expenditures in armaments exceeded US$997 billion, a sum more than 17 times the total contributions of OECD nations to aid designed to alleviate chronic poverty and disease. Due to continuing conflicts in Iraq, Afghanistan and elsewhere, investments in armaments will persistently rise while aid directed to reconstruction of Iraq and Afghanistan will overshadow international development as perceived by Pearson, Brandt and Brundtland.

UNDP Human Development recommendations

HDR 2003 recommends for people to be lifted out of poverty, governments and international agencies must:

   a.   invest in health care, education, and ensure access to clean water and hygienic sanitation

b. increase agricultural production and productivity

c. improve rural infrastructures

d. adopt policies to stimulate labour-intensive rural industries

e. ensure equity among all citizens, particularly between the sexes

f. legislate to ensure sustainable environments

g. promote competitive industries to process goods for export.

General comment

Poverty is both a cause and a result of insecurity: economic, food and health insecurity. Abject chronic poverty with consequent insecurity are fundamental causes of so-called terrorism. When millions of young people are denied access to employment with income sufficient to provide for their basic needs, when their territories are invaded, their property destroyed or taken from them by more powerful nations, they will inevitably resort to violence against their oppressors. 'Terrorism' will persist and become ever more violent so long as powerful nations, ignoring the edicts of the Atlantic Charter, seek to inflict their beliefs and concepts of governance on weaker folk by military force.

The Millennial objectives of reducing poverty by half are admirable but are of questionable realisation. This sorry state will persist so long as those with wealth and power are resistant to more modest and conservative consumption, to sharing their national and personal resources with people less fortunate both within and beyond their national boundaries. As Mahatma Gandhi so sagely observed: "The Planet's resources are sufficient to satisfy everyone's need but not everyone's greed".

The unwillingness of the most affluent to share their possessions with those less fortunate emphasises that redistribution of assets and more equitable access to resources and opportunities must be enacted by governments, encouraged by the people who elect them. In the wise words of the late Dr B.R. Sen, former Director General of FAO, "One person's poverty is every person's poverty; one person's need is every person's need."

6

Development in Agriculture and Biotechnologies

Sustainable survival

All living organisms are driven by two motivating forces: to reproduce themselves and to survive. Survival depends largely on protection from their enemies and having access to sufficient food. Consequently, for countless millennia the developmental history of hominids has to do with their search for sustenance and survival: an uneven progression from hunting and gathering to crop cultivation and livestock husbandry, the discovery and devising of biotechnologies by which to increase, protect, preserve and transform natural and cultivated biological species and materials into food, fibre and other goods of personal, domestic or commercial utility.

History records that agriculture has progressed as much through empirical perception as from scientific intention. In today's world, intensive large-scale highly mechanised production systems exist together with simple labour intensive subsistence systems where a few crops and farm animals survive on less than a half-hectare of land. Some nations and regions, which for many centuries were prosperous, maintained efficient profitable agricultural systems and produced well in excess of their needs, subsequently lost or forsook their agricultural capabilities and descended into a state of food insufficiency.

To be sustainable, agriculture requires a complex of critical resources, not least human skills and experience, and the dedicated political will of governments and those who legislate each nation's priorities. Libya and Nigeria are two nations where the discovery of oil has led to gross neglect of agriculture. In other ecologies, changes in weather patterns or excessive exploitation of natural resources and depredation of ecosystems have together degraded once fertile lands into near deserts.

Agriculture and biotechnologies

During the first century BCE, a Roman farmer called Varro is reported to have said: "Agriculture is a science which teaches us what crops should be planted in each kind of soil, and what operations are necessary in order that the land will produce the highest yields in perpetuity". The fourth century Bishop Ambrose of Milan wrote in *De Officiis* that agriculture is the only way of making money that gives no offence.

'Agriculture' is derived from the Greek word 'αγροζ' which means 'arable land', and 'αγρονομια' (English equivalent: 'agronomy'): the management of land. 'Agriculture' relates literally to the cultivation and management of land, but in popular usage embraces virtually all patterns and scales of crop and livestock production, including horticulture, (literally: the cultivation of gardens: Latin '*hortus*' = garden), more generally used to designate the cultivation of fruits and vegetables. 'Agroforestry' or 'Agri-sylviculture' relate to the integration of crop and livestock production with natural and cultivated arboreal species, in some instances the production and utilisation of non-timber forest products such as tree fruits, nuts, wild honey and medicinal plants. 'Agribusiness' has come to embrace most industries that provide goods, machinery and services to agriculture, and the industries that preserve, process, store, distribute and market the products from agriculture.

Agriculture, horticulture and biotechnologies are predominantly devoted to human health: the production of food to provide energy for healthy activity, bodily growth and restoration, drugs to diagnose, prevent and cure diseases, and fibres from which garments and fabrics are created.

'Biotechnology' is derived from several Greek words: 'βιοζ' meaning 'life', 'τεχνικοζ': 'an industrial art or craft', and 'τεχνολογια': 'systematic application of an industrial art or craft'. In recent literature, 'biotechnology' (singular) is frequently used exclusively to connote genetic modification of organisms and related modern biochemical reactions. In reality 'biotechnologies' (plural) embrace all processes that serve to protect, preserve, extract and convert materials of biological origin into products of industrial, commercial, social, economic or hygienic value and utility. Milling, baking and fermentation processes, devised by ancient Egyptians and Babylonians are as correctly designated 'biotechnologies' as are synthetic and extractive processes of biological materials that derive therapeutics and diagnostics from genetically modified organisms. One could cite instances of 'biotechnology' used to describe processes that have not progressed beyond a laboratory or field experiment, which more accurately should be designated 'biosciences'. In the subsequent text the plural noun 'biotechnologies' denotes the broader concept defined above.

Evolution of Planet Earth

The history of Planet Earth's existence can be classified geologically, according to the changing structure and composition of the earth's outer crust, or by progress in the development of tools, implements and technologies. Geological timing starts with the Precambrian period (about 4,500 million years ago), succeeded by the Cambrian (about 600 million years ago), which is believed to coincide with the appearance of the earliest fossils. Of particular interest to humans and their immediate ancestors is the Pleistocene period that began about 2 million BCE, during which period the Earth's climate underwent fluctuations between warm and cold temperatures, cycles that have continued until the present Holocene period that began around 15,000 BCE. The Holocene began with the end of the final glaciation, believed to have lasted from about 60,000 to 15,000 BCE. One of our many hominid ancestors, Neanderthal Man, a cave dweller and hunter, who appeared about 200,000 BCE, survived by hunting and gathering during the Pleistocene period.

From an agricultural perspective, an Earth-time classification based on the tools hominids used is perhaps more meaningful. It begins some 2.5 million years ago with the Early Palaeolithic period when hominids used primitive sharp-edged stone tools to harvest vegetation, cut up carcasses and sharpen sticks (Greek: Παλαιος = Ancient; λιθος = stone). During the Mesolithic period (Greek: μεζο = middle or between) that began about 15,000 BCE, stone tools came with multiple sharp edges, attached to handles fashioned out of wood or deer antlers. The Mesolithic era merged into the Neolithic (Greek: νεος = new) when more refined and efficient tools were made of polished stone, used as axe-heads to cut down trees, as sickles in crescent shapes with handles to harvest crops, with sharp blades to scrape animal hides. Harvested grains were ground by stone pestles and mortars, the pestle a heavy stone club, the mortar probably a concave hollow in a rock.

Hunting and gathering

Several scholarly narratives describe how our ancestors progressed from hunting and gathering to settled crop cultivation and animal husbandry [Blaxter 1992, Critchfield 1981]. Blaxter writes that to plan for the future we must build on knowledge accumulated from long past experience. The immediate past can provide a basis for simple extrapolation and adaptation of recent trends, while human history over the distant past can illustrate many of the difficulties that have confronted, and destructive practices that have constrained, humans in their quest for an adequate food supply. Blaxter discusses how changing weather patterns, massive epidemics, natural and man-made instabilities have been encountered and thus, in a broad sense, indicate the potential difficulties that lie ahead. Critchfield illustrates how settled agriculture, the practice of irrigation, domestication of animals and the invention of the plough led to the early establishment of village communities and their eventual expansion into towns and cities.

All animals in the wild survive by hunting and gathering, as, more than one million years ago, did the earliest hominids: the Australopithecines (African 'Ape-men'), the Pithecanthropoids distributed over Africa, Europe and Asia between 500,000 and 1,000,000 years ago. *Homo habilis* and the Neanderthals (*Homo neanderthalensis*) hunted and gathered from *ca* 200,000 to 35,000 BCE and devised Mousterian flint tools, more advanced and efficient than those of their predecessors. The first modern humans: *Homo sapiens* probably appeared between 10,000 and 15,000 years ago, being preceded by countless earlier primitive hominids. [McEvedy 1976; Weinberg 1994]

Anthropological opinions differ on when the various hominid species first appeared on earth and when some became extinct. *Homo erectus*, so named because of an apparent ability to walk upright, may have existed more than 1.5 million years ago. It is speculated that the Neanderthal hominids (*Homo neanderthalensis*) arrived about 200,000 years ago. Scientists at the University of Utah have studied evolution of certain head lice in tracing the history of early hominids. They conclude that *Homo neanderthalensis* and early *Homo sapiens* lived contemporaneously until about 25,000 years ago, when the Neanderthals and some other early hominids disappeared. The Utah scientists conclude that *Homo sapiens*, possessing larger brains and possibly bigger bodies may have destroyed their more primitive hominid cohabitants [Henderson 2004].

Hunting and gathering persists as a source of sustenance for many primitive people. Survival by hunting and gathering is sustainable if the dependent humans do not increase in numbers faster than the natural rate of food source regeneration. It is speculated that as each hunting-gathering tribe exhausted its immediate natural food resources it would migrate to find other sources of supply [Blaxter 1992]. Whenever a population grew faster than the natural regeneration of the wild animals and plants that provided its food some or all would die of starvation. It is not improbable, as has been the case in recent times, that the least productive, young children and old men and women, would be the first to be sacrificed in times of shortage.

Early hominid diets

The ways in which prehistoric hominids acquired their food and the composition of their various diets can only be conjectured, accurate determinations being virtually impossible. To begin, sites for suitable study must be found. In dry environments biological remains are better preserved, so more sites have been studied in arid and semi-arid ecologies than in the humid tropics. Samples suitable for analysis are plant and animal parts that have resisted decay for several thousand years. Fleshy, non-lignified plant parts rarely survive for any length of time. Carbonised seeds and highly lignified seed coats are among the best survivors. Bones of very small animals and of fish are less resistant to degradation than large animal bones and mollusc shells. Consequently, what can be analysed illustrates only an unknown proportion of the foods consumed by ancient hominids [Blaxter 1992; Verpmann 1973; Wheeler *et al* 1989].

Palaeoclimatology, carbon dating, pollen analyses, isotopic and elemental analyses of hominid skeletal remains are methods variously employed to discover some of the components of ancient hominid diets. The ratio of ^{13}C to ^{12}C in collagen from hominid bones indicates plant and animal sources that were dominant in the hominid's diet [Price 1989]. The ratio of ^{15}N to ^{14}N helps to distinguish between leguminous and non-leguminous diets. Where written or pictorial records are available, as from the early riverine civilisations of Egypt, Mesopotamia and the Indus valley, dates and patterns of the transition from hunting-gathering to settled farming can be more reliably determined. Nonetheless, present analytical methods cannot distinguish between plant and animal remains from a wild or a domesticated species. Evidence of lined storage pits and stone pestles and mortars do not irrefutably indicate that the grains stored and processed were cultivated, since both could be used for grains harvested from the wild. Out of the many animals slaughtered by hunters, only a few species appear to have been domesticated. Studies and samples analysed from the Mesolithic and early Neolithic periods show that wild red deer and roe deer were abundant, though neither seem to have been domesticated. [Blaxter 1992; Wheeler & Jones 1989; Clutton-Brock 1987].

Middle Palaeolithic hominids, whose remains are spread across Europe, West Asia, the Middle East and parts of Africa, were more successful hunters than their predecessors. Animal remains suggest they devoured many species of large mammals; how much of their meat came from hunting and killing live animals, how much from the scavenging of dead carcasses is unknown. Sites in Europe where bones from several animal species were found may have been primitive abattoirs where hominids processed the spoils of their kill. Archaeologists have discovered sharpened pointed stones probably used as spear heads, indicating a new kind of hunting technology. Simple hearths at many Middle Palaeolithic sites show that fire was used to cook meats, to dry and cook grains and, with protective clothing from animal hides, enabled Neanderthals to spread to and survive in such colder regions as the Central Asian steppes.

The Mesolithic period extended from the end of the Pleistocene ice age until the Neolithic, about 10,000 years ago, when settled farming first appeared in different locations and at different times. The end of the Pleistocene ice age led to rapid environmental change where warmer post-glacial conditions of the Holocene epoch caused ice sheets to retreat and sea levels to rise. Temperate forests spread across Europe and Asia, large herds of mammals such as reindeer were replaced by red deer, roe deer and wild pig. Reindeer, elk and bison, adapted to cold climates, retreated to the north; the mammoth, giant deer and woolly rhinoceros became extinct.

From hunting to settled agriculture

Settled agriculture must be regarded as among the most important innovations and developments in human history. For whatever reasons particular communities changed from hunting-gathering to agricultural technologies devised and developed from perceptive empiricism, each adopted system being influenced by local soil, climate and resource conditions. There seems general agreement among archaeologists and ethno-botanists that transitions from hunting and gathering to settled farming occurred in different ways, according to different patterns and stimuli in different places. Settled agriculture, in which food-crop seeds were planted and animals domesticated began during the Neolithic period around 10,000 years ago. In some communities settled farming seemed to result from local indigenous development, in others farming technologies devised or discovered in one location radiated outwards to be adopted or adapted by other communities [Blaxter 1992,Critchfield 1982].

To evolve from gathering to crop cultivation required the discovery of indehiscent plant species. In order to reproduce themselves, wild grasses and other plants when mature scatter their seeds, some of which will then take root. Seed scattering plants are described as dehiscent (= bursting or splitting open). This trait of dehiscence is particularly noticeable in sesame, hence the expression 'Open sesame'. Settled cultivators have need of indehiscent plants so that, when mature, the seeds can be harvested and stored, to be eaten or later planted.

The earliest farmers were of Neolithic culture located in the region now covered by Iraq, Iran, Jordan, Syria and Turkey, along the Nile in Egypt, the Indus river valley and its tributaries in what is now India and Pakistan. Exactly when the first crops were derived from planted seed is uncertain and varied among regions. Though most are believed to predate 6000 BCE, some ethno-botanists believe the earliest may date from around 10,000 BCE. According to carbon isotope dating, wheat, barley and possibly flax were cultivated in Mesopotamia from 8,000 BCE. Wheat and barley cultivation slowly spread northwest to reach Britain about 2,000 BCE. A diversity of fruits and vegetables:

onions, melons, cucumbers, dates, figs were cultivated in Mesopotamia and nearby regions between the fourth and third millennia. During the second millennium, cotton was grown and spun in India and Egypt, linen and silk in China.

Examination of animal and plant remains by ^{14}C indicates that sheep, bovines, goats, pigs and asses were domesticated in Mesopotamia between 9,000 and 6,000 BCE. Neolithic settlements were more permanent than the camps of nomadic hunters and gatherers though some early farming settlements had to move periodically if weather patterns changed or soils became infertile from excessive cultivation [Blaxter 1992].

The first crop cultivation systems probably began by clearing land for continuous cropping until soil fertility was exhausted, at which time the cultivators would move to another location, which then and now in certain poor nations is the basis of slash-and-burn agriculture. It is generally assumed that the early cultivators started by empirical discovery of which wild plants were edible and from which seed could be removed and planted on cleared land. From the harvest of their cultivated crop, farmers learned to save some seed for the next planting. Repeated cultivation of a chosen genotype would eventually provide the farmer with a stable crop with relatively homogeneous characteristics.   Herds of goats and sheep were assembled, captured young wild animals with the most useful characteristics, those with small horns and high milk yields, being domesticated. Their progeny would become progressively tame, with comparatively uniform physiologies.

Demographic, economic and social change: the river valley civilisations

Settled farming, with domestication of plants and animals led to profound social changes, particularly in regions such as Mesopotamia, Egypt and the Indian Harappan culture, where fertile soil, ample water for crop irrigation and abundant sunlight provided conditions for highly productive farming systems. Archaeologists have located early farming sites in Thailand and China, where rice and other indigenous crops were cultivated, in a few African territories where indigenous sorghum and millets were grown. Most comprehensively described were farm settlements along the Euphrates, Tigris, Nile and Indus valleys [Critchfield 1981; Thapar 1996].

The capacity to produce food surplus to the producers' needs, facilitated trade in agricultural commodities, the establishment and growth of non-farming urban communities. It also led to income stratification as prosperous farmers and urban traders increased their wealth. Cultivators in Mesopotamia and Egypt were probably the first to use systematic irrigation, to establish cities, towns and regulated social organisations. The Indian civilisations were not far behind. The development of cuneiform script in Mesopotamia and Egyptian hieroglyphics [the writing of priests] provided means of communication and records for posterity [Leick 2002; de Burgh 1947].

Mesopotamia

'Mesopotamia' means 'between rivers' [Greek: μεζο = 'between'; ποταμοζ = 'river'], the rivers being the Tigris and Euphrates that rise in the Anatolian mountains and flow south to the Persian Gulf. In contemporary geographical usage, Mesopotamia includes Iraq, eastern Syria, parts of Iran and much of south-eastern Turkey. It comprised two distinct ecological zones: the northern Fertile Crescent, where rain-fed and irrigated agriculture started and the first urban settlements appeared some 10,000 years ago; and the southern flood plain, the rich fertile land where, in dry seasons, crops were irrigated from an ingenious system of dykes and canals. The most enduring legacies of ancient Mesopotamia are the dozens of established cities, and the invention of writing. Each of the cities strung along the rivers controlled its own rural territories, its agricultural production and system of irrigation, its individual form of governance and administration.

Many of the ten principal cities – starting with Eridu about 8,000 years ago and on to Babylon, which around 2500 BCE grew from a small town to a magnificent city that lasted into the first centuries of the contemporary era – disappeared over time variously because of climatic change, rivers changing their courses, military conflict and reasons not fully understood. Most now exist only as archaeological remains [Leick 2002].

According to some historians, most of Mesopotamia was under water prior to the Chalcolithic period. This is possibly the origin of the Biblical flood mythology. There is an ancient Sumerian myth that the storm god Enlil decided to reduce the population by drowning them in a flood. The god of wisdom, Enki, persuaded a wise man Atrahasis (also known as Ziusudra) to build a boat and to stock it with seeds and species of all living organisms, so as to ensure the survival of all creatures when the flood subsided [Armstrong, K 2005].

As the climate gradually changed, some 6,000 years ago, the floods receded, and the plains became habitable. The post-deluvian virgin alluvial soil supported barley, wheat, flax, horticultural crops and livestock production. The southern marshlands were a substantial source of fish, game and wild life. With water from the Tigris, Euphrates and their tributaries, crops were irrigated during seasons of low rainfall. Thus, starting in the fourth millennium, 4000 to 3000 BCE, southern Mesopotamia [Babylonia, Akkad and Sumer] provided an extensive exploitable ecosystem able to produce abundant food, thus encouraging settlements which expanded into large urban areas. Produce from farmers, fisher-folk and hunters stimulated a transition from subsistence agriculture to rural-urban complexes. Donkeys and oxen were probably the main beasts of burden.

Urban commerce

The urban traders, artisans and administrators were fed by abundant food from the surrounding cultivated land. Artisans developed pottery from native clays, textiles out of flax and wool of domesticated sheep. Administrators prescribed and enforced rules that restrained conflict, enabling people spread among the settlements to live together in comparative peace. Surplus grain, leather from sheep hides, sun-dried fish, dates, other fruits, textiles and timber from natural forests, and locally mined metals and semi-precious stones were exported to urban settlements along the navigable Tigris and Euphrates [Leick 2002].

Theophrastus, Aristotle's celebrated pupil, in his *History of Plants*, describes how in ancient Babylon wheat fields were harvested twice each year, beasts being subsequently fed on the residual stubble. He stated that on cultivated, irrigated land each planted wheat seed could produce between a 50 and 100-fold harvest. One assumes he was comparing the total weight of biomass harvested not the weight of harvested seed to the weight of the seed planted. On fertile soil irrigated from nearby rivers, Babylonian farmers could harvest two crops every year, sufficient to satisfy their own needs and those of the expanding urban populations. [de Burgh 1947].

Egypt and the Nile

The Greek historian Herodotus wrote: "Egypt is the gods' gift of the Nile". Through Egypt, the Nile flows some 1,200 km from the first cataract in the north to the broad delta where it empties into the Mediterranean. The Egyptian Nile is fed from the White Nile, which originates in northern Uganda, supplemented by the Blue Nile, which flows in at Khartoum, and the Atbara which joins about 450 km north of Khartoum. Both the Blue Nile and the Atbara originate in the Ethiopian highlands. Over many centuries the abundance of the Nile, the fertile silt deposited after each annual flood, and Egyptians skilful management of dykes and irrigation systems supported a flourishing agrarian society. Diverse plant species were cultivated for food, clothing and shelter, well into the period when, under Roman rule, Egypt was regarded as the breadbasket of the Mediterranean [de Burgh 1947; Livy 1960; Cowell 1948]. The Nile sustained Egypt's fishing industry for several thousand years.

The Egyptian population grew from several hundred thousand during the pre-dynastic period (5000–3000 BCE) to several million by 1000

BCE. Egypt's cities and settlements were located close to the Nile, the densest population being along the flood plains. During most years when the rains in Uganda and the Ethiopian highlands were abundant, the Nile overflowed its banks. The floodwaters as they receded deposited a rich, black fertile soil to sustain productive crop cultivations over many millennia [McEvedy 1976].

History records the decline and fall of many once prosperous and advanced nations. Egyptians were literate, numerate and practised productive irrigated agriculture more than 5,000 years ago. They devised a decimal numerical system and the means to communicate in writing. The mathematics of dynastic Egypt was significantly influenced by agriculture. Calculations of area, length x breadth, were used to determine the size of farmers' fields, and before the fourth Millennium Egyptians calculated the value of π [pi] the ratio of circumference to diameter as equal to 3.145. The present accepted value is 3.143. Knowing the value of π enabled Egyptians to calculate the storage capacity of cylindrical grain bins.

Indus and Harappan culture

The earliest traces of hominid activity in India go back to the second interglacial period between 400,000 and 200,000 BCE and show evidence of stone tools. There followed a long period of slow evolution that culminated in the spectacular Indus valley civilisation, known as the Harappan culture, which matured around 2500 BCE. The antecedents of the Harappan culture were village sites in the Baluchistan hills and rural communities along rivers that flowed in Rajasthan and Punjab. The Vedic civilisation, which began in the Indus valley, is said to have believed more in monuments of the mind and thought than in material monuments. Archaeological excavations have brought to light some 80 settlement sites dating from about 3500 BCE. The Indus civilisation was among the world's earliest; Indian historians wisely state that civilisation began when cereals and cattle were first domesticated. The Indus civilisation culminated in two main centres: Mohenjodaro close to the Indus valley, and Harappa on the Ravi river. Wheat cultivated near to Mohenjodaro is claimed to be an ancestor of the wheat varieties now grown in the Punjab. Both Mohenjodaro and Harappa maintained large granaries in which the annual wheat harvest was stored. In India cotton was first grown. Grain, cotton and other commodities were transported in wheeled carts drawn by oxen, and by river boats [Mookerji 1961].

The Harappan culture was essentially urban, the two main cities Mohenjodaro and Harappa being maintained from surplus crops produced by the surrounding rural farmers and stored in elaborate granaries. Harappan communities developed a flourishing trade with people in the Persian Gulf and Mesopotamia [Thapar 1966]. Some historians believe the first Sumerians to settle in Mesopotamia migrated from Baluchistan [Bouquet 1962].

Roman agriculture

Marcus Tullius Cicero, Roman Consul, politician and philosopher, wrote: "Of all gainful occupations nothing is more pleasing and better becomes a well-bred man than agriculture" [Cowell 1948]. Some 6,000 years ago active volcanoes in Central Italy covered the Latin plain with ash and lava that provided Roman farmers with an extremely fertile soil. For several centuries Roman farmers enjoyed exceptional prosperity. Between 265 and 70 BCE the number of male Roman citizens rose from 290,000 to 900,000. When women, slaves and non-citizens were counted, during the last century BCE, the total Roman population probably exceeded 3 million [Tacitus 1965]. But soil fertility and productivity could not keep pace with population growth, a constraint aggravated by top-soil stripped of its protective tree and grass cover being washed away by torrential rains. Animal manure and wood ash were insufficient to replace lost soil nutrients and what was once fertile crop land degenerated into land fit only for grazing animals. To pursue Rome's military adventures and conquests, many of the best farmers were recruited into the Roman legions. By the turn of the millennium, as BCE changed to CE, Rome imported about 190,000 tonnes of wheat from Sicily, Sardinia and North Africa. Soon thereafter Egypt, the principal granary of the Mediterranean, became the main supplier of food grains to Republican Rome [Cowell 1956; Livy 1960].

Of the three civilisations whose early prosperity was derived from productive agriculture, only India remains as an efficient producer of food grains. As recorded in *Indian Agriculture 2003*, published by the Indian Economic Data Research Centre, since the 1950s India's annual wheat harvest has risen from 6 million to 70 million tonnes.

From manual to mechanised processes

Virtually all the technologies employed to provide for basic human needs – agriculture, food processing, drugs, pottery and textiles – were discovered and developed empirically. Domestic and artisanal technologies were practised for many centuries before there was any scientific understanding of why they worked. Scientific understanding began in the late eighteenth and nineteenth centuries; only during the last century have novel biotechnologies been devised and developed based on scientific principles. For many centuries most apparent progress was from gradual, episodic mechanisation: the replacement of human hands by machines. Long ago iron replaced wood in simple ploughs, the primitive iron plough, stronger than wood, turned top-soil over rather than just scratching the surface. Kates [1994] illustrates how new tools and technologies have stimulated human population growth over the past million years by raising available food supplies. Progressive improvement in hand tools used by hunters and gatherers resulted in an estimated population increase from 150,000 to 5 million during the period from one million to 100,000 years ago; a second surge from 5 million to *ca* 1 billion came with expansion of settled agriculture between *ca* 8000 BCE and the middle of the nineteenth century, since when the application of science to food production, preservation, processing and distribution has stimulated a world population explosion to more than 6 billion [Kates 1994; Hulse 1995].

Most rapid and far-reaching farm mechanisation spread across the Western Great Plains of the United States during the nineteenth century. The steel blade ploughshare appeared in 1830, the reaper in 1834, soon followed by mechanised harrows, seeders, threshers and binders, many manufactured and marketed by Cyrus McCormick. Mechanisation disruptively reduced the need for human labour on American farms: one McCormick threshing machine operated by four men replaced 400 farm labourers. Towards the end of the nineteenth century, guano from Peru and nitrates from Chile began to replace or supplement animal manure as nitrogen fertilisers. Introduction of the steam tractor accelerated farm mechanisation. During the second half of the nineteenth century more than 160 million hectares of new land came under cultivation in the United States [Nye and Morpurgo 1955].

As in ancient Egypt and Mesopotamia, rising farm productivity stimulated the growth of cities, the rate of American city expansion being unprecedented in human history. In 1910 Chicago's population reached 20 times the size in 1860. Mechanisation, more abundant farm productivity and rural-to-urban migration occurred in Europe and North America. During the twentieth century, the proportion of the Canadian work force employed on farms fell from *ca* 40 per cent to barely 3 per cent. The human misery suffered in cities that grew too rapidly, without planning or provision of essential services, is forgotten or ignored among nations of Africa, Asia and Latin America where urban expansion is uncontrolled, where slums and unhygienic environments are all too familiar.

From empiricism to scientific agriculture

One of India's most distinguished scientists, the late Professor Y. Nayudamma, made the profound observation that science and technologies serve no useful purpose if they fail to provide for the needs of human beings. India's agricultural scientists and farmers in productive States have served its peoples exceptionally well, transforming the nation from dependence on imported grains and food aid to an abundant self-sufficiency in food grains.

Contrary to long-standing belief, agriculture does not inevitably progress starting with scientific research that begets invention, that leads to technological innovation, and ultimately results in ever-increasing crop production. Farming systems research (FSR), that starts by assessing and understanding farmers' adopted technologies, opportunities, constraints, resources and risks, began during the early 1970s. The benefits to farmers realised by FSR have been less widely publicised than research that generated higher yielding cereal crop phenotypes. It is through modifications based on careful systematic analysis and understanding of existing farming practices that sustainable systems are most effectively and sustainably developed, adapted and adopted. [IDRC 1981; Zandstra *et al* 1981].

Blaxter contends that modern food systems, highly dependent on energy other than solar radiations, are inherently unsustainable. As has come about over the past 12,000 years, novel or improved technologies and systems are needed to maintain an equitable, sustainable balance between food supply and persistently rising demand [Blaxter 1992]. Blaxter estimates that roughly 98 per cent of today's world's food supply is derived from terrestrial ecosystems. Though some 80,000 identified plants and countless animal species are believed to be edible, roughly 90 per cent of all human food calories come from only 15 plants and eight animal species.

Early scientific evolution

In his imaginative novel *Brave New World* Aldous Huxley asserted that the world's history of economic, technological and industrial development can be divided into two periods: a. before and b. after Henry Ford, the Detroit automobile industrialist who devised the continuous on-line process by which to assemble automobiles. In this author's opinion a more rational division is pre- and post-Faraday. If the Uruk-Sumerian invention of cuneiform script in the fourth Millennium, and Johannes Gutenberg's creation of the printing press in the 1450s can be considered the most important contributions to international scholarship, it seems appropriate that Michael Faraday started his working life as a bookbinder. Though he never graduated from a university, Faraday's fascination with, and self-acquired knowledge of science led to his appointment as scientific assistant to Sir Humphrey Davy. Faraday's renowned contribution to electromagnetism began when he reviewed the Dane Hans Christian Oersted's discovery of the magnetic effect associated with an electric current. Faraday was the first person to discover benzene whose six-carbon ring structure, later elucidated by Kekule, is of fundamental importance to organic chemistry, to agricultural and food biochemistry [Schumm 1987].

Science together with empiricism

Although during the fourth century BCE Aristotle classified more than 500 animal species within hierarchies, and his pupil Theophrastus wrote a *History of Plants*, agricultural practices changed little from the time men and women began to cultivate crops and husband livestock around 10,000 years ago. Thales, one of the first Greek philosophers in the sixth century BCE, postulated that water was the main constituent of all plants. But it was not until the seventeenth century CE when van Helmont observed and reported the dependence on water of his experimental willow, Stephen Hales demonstrated how water is transmitted through roots and leaves, Joseph Priestley and Jan Ingenhosz described leaf respiration, and von Liebig and Dumas experimented with plant uptake of nitrogen and other nutrients, that an understanding of what influences plant growth began to emerge [Mees 1947; Gribbin 2002].

From the time of Faraday and his contemporaries in the nineteenth century scientific explanations of natural phenomena, of technologies empirically discovered, devised and developed over many millennia, gradually emerged. Nonetheless, while scientific underpinnings were coming to light, discovery and progress from thoughtful observation was widely evident. During the eighteenth century an English farmer,

Coke of Okeham, improved soil fertility with little help from chemists; while Robert Bakewell increased meat and wool per animal by selection and breeding of Leicestershire sheep more than 100 years before Mendel discovered the laws of heredity. The famous Rothamstead experimental station blossomed out of field experiments started by a private farmer John Lawes, who bore the total cost for almost 60 years until in 1889 a sustaining Trust Fund was endowed to carry on some of the world's longest systematic studies of soil fertility. During the seventeenth century, Grew, Camerarius and Sprengel described plant male and female organs – stamens, styles, ovaries and ovules – and how insects carried pollen between plants.

Microscopy and cytology

Who first discovered the magnifying power of convex lenses is unknown. The thirteenth century English scholar Roger Bacon is said to have constructed eye spectacles. During the seventeenth century Galileo devised a compound microscope, equipped with convex lenses he had ground. In 1625 he published microscopic details of various insects. In the late seventeenth century, a self-educated Dutch microscopist Anton van Leeuwenhoek, who earned his living in a drapery store, delicately ground glass lenses used to study and describe previously invisible microscopic materials. He described tiny parasites that infested fleas, a discovery that inspired Jonathan Swift to write:

> So naturalists observe, a flea
> Has smaller fleas that on him prey;
> These have smaller ones to bite 'em
> And so proceed *ad infinitum.*

Van Leeuwenhoek described other cellular organisms that he named 'animalcules', which probably were protozoa or bacteria. During the second half of the nineteenth century, bacteria were more accurately identified by Louis Pasteur, who demonstrated how micro-organisms could be thermally inactivated.

From alchemy to chemistry

Robert Boyle, son of the Earl of Cork, is most renowned for demonstrating the inverse relation between gaseous volume and pressure. Whether or not his book, *The Sceptical Chymist: Chymic-physical doubts and paradoxes*, provided the bridge between alchemy and chemistry, Boyle undoubtedly pioneered scientific method, inspired by Francis Bacon's thesis that scientific investigations must begin by critical assessment of all relevant available data, not by imagining a wonderful idea then seeking facts to sustain it. From the time of Faraday, progress in scientific method, in analytical chemistry, microscopy and cytology progressed to a greater understanding of living organisms, development of agricultural, food and pharmaceutical technologies.

The quarter-century following Faraday's empirical discoveries of electro-magnetic induction and the anaesthetising properties of ether was a vintage period in scientific progress. Darwin published *The Origin of Species*, Mendel formulated his laws of genetic inheritance, Maxwell published his theory of electromagnetism, Pasteur and Koch described the microbial origins of many diseases, Kekule elucidated the cyclical structure of benzene, Gibbs published his first laws of thermodynamics related to chemical reactions, Geissler's vacuum-tube technology paved the way for discovery of electrons ['cathode rays'], X-rays and the eventual discovery of radioactivity. A congress at Karlsruhe, convened by Kekule, led to a rational nomenclature for chemical compounds. The classification of elements according to their atomic weights by Cannizzaro, Newlands, de Chancourtoix and Meyer culminated in Mendeleyev's periodic table of all known elements. The later classification by atomic number did not greatly alter the Medeleyev table.

Joseph Black, Henry Cavendish, Joseph Priestley and Carl Scheele advanced the development of chemistry as a distinct science. Their discoveries of carbon dioxide, oxygen, hydrogen, ammonia, hydrogen chloride and sulphur dioxide; of the composition of water; and their finding that air is a mixture of gases, are illustrative of what was achieved by men of immensely diverse interests, most of whom, by contemporary scientific standards, would be classed as non-professionals [Gribbin 2002]. An attempted dissertation of all that science and technology have contributed to modern agriculture and biotechnologies lies far beyond the scope of this publication. Nevertheless, a few exceptional persons deserve special mention.

Lavoisier

Antoine-Laurent Lavoisier was born in 1743 in Paris, the son of a well-to-do lawyer. Lavoisier qualified as a Licenciate of Law, at the same time studying astronomy, botany, chemistry, geology and mathematics. He never practised law, his first work being a project to complete a geological map of France. During the late eighteenth century, Lavoisier devoted his talents to chemistry, demonstrating that when sulphur burns it gains weight, the first step in understanding how atmospheric oxygen (Priestley's 'pure air') is critical to combustion processes. From later experiments he concluded that animals derive their body heat from combustion processes dependent on inspired oxygen. He determined the weight of ice melted during ten hours by the heat from a guinea pig's body. Then he measured the weight of carbon burned to melt the same weight of ice. He measured how much 'fixed air' or CO_2 the guinea pig exhaled during ten hours and how much 'fixed air' was generated by burning different weights of charcoal. Lavoisier thus anticipated von Liebig's demonstration, 100 years later, that animals derive their body heat and energy from *in vivo* conversion of carbohydrates and lipids.

Lavoisier was the first to demonstrate that water is formed by the combination of hydrogen ('inflammable air') with oxygen, and how 'fixed air' is a compound of carbon and oxygen. His books *Traité elementaire de chimie* and *Méthode de nomenclature chimique*, published in 1787, laid the foundation for chemistry as a distinct scientific discipline, and provided the basis for a logical and systematic nomenclature for chemical elements and compounds: 'oxygen' replacing the earlier 'phlogisticated air', 'inflammable gas' giving way to 'hydrogen', 'oil of vitriol' being renamed sulphuric acid. One-third of the latter book was a dictionary of new chemical names presented in alphabetical order together with the old names, many coined by alchemists on the basis of the colour and physical appearance of the substance [Gribbin 2002; Crosland 1978; Ward and Dubos 1972].

Lavoisier's many profound pronouncements include: "La vie est une fonction chimique", demonstrably true in that all life processes are sustained by the conversion of one form of energy into another by biochemistry; and "Scientists should be as precise and meticulous in their written and spoken words as in their experimental measurements", advice all too often ignored in speeches and writings about 'sustainable development'. Due to an earlier involvement with Tax Farmers, in 1794 the French revolutionaries decreed that Lavoisier be executed by the guillotine.

Chemical nomenclature

In 1806 Berzelius defined Organic Chemistry as: "That which describes the composition of living bodies and the chemical processes within them", the first distinction being between substances of vegetable and animal origin. Benefiting from Lavoisier's earlier publications, in 1860 Kekulé convened a congress in Karlsruhe at which a group of distinguished scientists discussed systems of nomenclature and symbols for organic compounds. From this congress and subsequent discussions Organic Chemistry progressed from Berzelius' definition to the present concept: the complex chemistry of thousands of carbon compounds [Crosland 1978; Dixon 1973]. It was not until the twentieth century that Berzelius' concept of organic chemistry evolved into the discrete discipline 'biochemistry', at first designated 'physiological chemistry'. From a typical dictionary definition: "Biochemistry is the study of the chemical constituents of living organisms, their functions, reactions and transformations" [Rose 1970].

Taxonomic classification

Ray and Linnaeus

Arguably, the most productive naturalist of the seventeenth century was an Englishman, John Ray. Ray travelled extensively throughout Europe observing, describing and collecting thousands of specimens of natural flora and fauna. The results of his studies are reported in his books: *Ornithology*, *History of fishes* and *History of plants*. The latter, published in three volumes between 1686 and 1704, describes more than 18,000 plants classified according to distribution, natural habitat and those believed to possess pharmacological properties. Ray established 'species' as the basic units of taxonomic classification. His book "History of insects" was published posthumously.

Ray's taxonomy was based on anatomy, morphology, physiology and systematic principles adopted by the better known Carolus Linnaeus, born in 1707 in Sweden [Gribbin 2002]. The Linnaeus family name, Ingemarsson, was legally changed by his father to Linnaeus, the Latinised version of the Linden tree. Though he graduated in medicine, Linnaeus' life-long interest was in biology. He noted differences between the reproductive organs of flowering plants as a criterion of classification. He lived until 1778, switching from the Chair of Medicine to the Chair of Botany at Uppsala University in 1742. His most famous book, *Systema natural*, was first published in 1740. His binomial system of naming species was spelled out in Volume 1 of the tenth edition in 1758. Linnaeus classified close to 8,000 plant species and some 4,400 animals, all arranged in a hierarchy from 'Kingdom' at the top to 'species' at the bottom. As taxonomic techniques were refined, some of Linnaeus' original Latin binomials have changed; but his principles remain as the standard of species identification: a Latinised generic plus a specific name: e.g., *Canis lupus*, the scientific name for 'wolf'.

Linnaeus was the first to include humans in a zoological classification system. He questioned why humans are the sole species '*sapiens*' in the genus '*Homo*'. In his 1746 publication *Fauna Svecica* he writes: "I have yet to find any characteristics by which man can be distinguished on scientific principles from an ape." Based on modern DNA evidence man should be classified among the chimpanzees, possibly as '*Pan sapiens*'. The accepted classification '*Homo sapiens*' is used so as not to offend certain religious sensitivities [Ward and Dubos 1972; Gribbin 2002]. As modified by later taxonomists Linnaeus classification of animals now reads:

Kingdom	*Animalia*
Phylum	*Chordata*
Subphylum	*Vertebrata*
Class	*Mammalia*
Order	*Primates*
Family	*Hominidae*
Genus	*Homo*
Species	*Sapiens*

After Linnaeus died, James Smith, a wealthy English botanist, was instrumental in 1788 in founding the Linnean Society which, in London, holds most of Linnaeus' collections of biological specimens [Gribbin 2002].

Genetic inheritance

The years in the nineteenth century when Faraday lived (he died in 1867) yielded an abundance of discoveries that advanced the development of agricultural science. In 1837, Schleiden proposed that all plant tissues are composed of cells. A year later, Schwann suggested that all living organisms are of cellular composition. In 1858 Virchow, a professor of pathology in Berlin, asserted that all cells are derived by division of pre-existent cells. Using more powerful microscopes, Hertwig and others later discovered cell nuclei and, working with sea-urchins, observed how sperm penetrated into eggs, the nuclei of the sperm and egg cells fusing into a new nucleus that combined material from both parent cells.

Flemming and van Beneden reported the presence of thread-like structures in cell nuclei, which they named chromosomes because of their susceptibility to staining by chemical dyes. At Freiburg, Weismann described how when a cell divides its chromosomes are duplicated to be shared between the two daughter cells. Weismann declared chromosomes to be carriers of hereditary characteristics between parents and progeny (Hulse 2004).

Mendel

Now recognised as the founder of classical plant breeding, Mendel was born in 1822 into a farming family in Moravia (then in the Austrian empire). Being too poor to afford a university education, Mendel joined the priesthood where he changed his Christian name from Johann to Gregor. After completing his theological studies, recognising Mendel's exceptional intellect and intelligence, the abbot of his monastery in Brünn (now Brno in the Czech Republic) enabled Mendel to study at the University of Vienna. There he acquired a knowledge of physics, chemistry, statistics, probability mathematics and plant physiology.

After two years, and without a degree, Mendel returned to the monastery where, in 1856, he began a ten-year study of heredity in peas. Working with several thousand plants, on a plot 35 × 7 metres, Mendel transferred pollen by hand from one plant to another, meticulously recording his every action and the results. His results from hundreds of repeated experiments, all subjected to rigorous statistical methods learned in Vienna, revealed that when peas with rough wrinkled seed coats were crossed with smooth-coated seeds, 75 per cent of the progeny had smooth coats, 25 per cent were rough, wrinkled seeds. This phenotypic phenomenon of unequal inheritance of parental characteristics illustrated the now familiar principle of dominant and recessive genes. Contrary to Darwin's theory of heredity, Mendel demonstrated that inheritance results not in a uniform blending of parental characteristics but from specific traits being inherited in unequal proportions.

Mendel's unique results were overlooked for more than half a century, in large part because in 1868 his abbot died and Mendel was appointed as the abbot's successor. His religious and administrative responsibilities allowed time neither to continue his plant breeding experiments nor to promulgate and disseminate his results to many who could have benefited. The rediscovery early in the twentieth century of Mendelian laws of inheritance, together with the elucidation of chromosomal behaviour, laid the foundations for highly productive plant breeding, most notably the so-called Green Revolution during the latter half of the twentieth century. [Stern and Sherwood 1966; Stoskopf 1993; Wortman and Cummings 1978; Gribbin 2002].

Closing comment

The foregoing illustrates how biotechnologies were discovered, devised and developed over many millennia; how the basic biotechnologies – agriculture, food, drugs and textiles – began as domestic or artisanal crafts and gradually, over many millennia, were modified by perceptive empiricism. Early, primitive technologies preceded the most rudimentary scientific understanding by many hundreds of years, systematic scientific discovery being of very recent origin with much yet to be learned. What is known of the development of biotechnologies essential to human health and survival: agriculture, food, drugs and textiles, can truly be described as 'a gradual unfolding' and illustrate that sustainable progress can rarely be hastened but more reliably results from persistent patient systematic imaginative activity.

As is emphasised later, all development – to be rational, productive and sustainable – requires a systematic understanding and logical integration of all that influences and impinges upon the intended development. This is especially true of development related to the biotechnologies, upon which human life, biodiversity and their environments depend, biotechnologies which are progressing and changing more rapidly than at any time in the past.

7

Sustainable Agriculture

Diverse and divergent concepts

Recent years have witnessed an immense outpouring of published literature under the general heading of 'Sustainable Agriculture'. In preparing a report on "Sustainable agriculture and food security" for the 1992 UN Conference on Environment and Development a scientific Commission of the International Council of Scientific Unions (ICSU) reviewed some 300 related publications [ICSU-CASAFA 1991]. The Commission reported how 'sustainable agriculture' is treated from a diversity of scientific, pseudo-scientific, biological, ecological, ideological and philosophical perspectives. The following are a sample of relevant significant publications referred to in the Commission's report: Agriculture Canada [1989], Altieri [1983], Boeringer [1980], Brown [1987], Brundtland [1987b], Bunting [1987], CGIAR/TAC [1989], Edwards et al [1990], FAO [1984], Harwood [1990], Jain [1983], NRC/NAS [1989], NABC [1989], Rodale [1983], Reganold and Papendick [1990], Swaminathan and Sinha [1986]. During the succeeding decade many more have been added.

As defined by the Brundtland panel, a sustainable agricultural system manages essential resources so as to satisfy the needs of all people presently dependent without compromising the needs of future generations. It is when authors try to prescribe agricultural production systems claimed to be of global ecological adaptability that confusions and contradictions arise. Geological and anthropological evidence indicate that, from the time of creation, the earth, its environment and inhabitant creatures have changed, sometimes gradually and evolutionary, sometimes in extreme and disruptive patterns. Many scientists contend that human activities are accelerating environmental and climatic change, some attributable to developments in agricultural technologies. The ICSU-CASAFA report illustrates how the relevant literature presents divergent opinions and recommendations based on experience with different locations, ecologies, technologies, environmental, climatic, social and economic circumstances. Sustainable systems are prescribed by commentators of variable international experience, some seemingly motivated more by ideological persuasions than from a logical, systematic scientific perspective.

Systems and technologies proposed for any farming community should be assessed in relevant context and require a comprehensive analytical understanding of the existing technologies, resources, constraints, prevailing ecological and climatic conditions; farmers' individual and collective capacities to tolerate the inevitable risk of adopting unfamiliar crop types or systems of cultivation. Farmers who produce surplus to their families' subsistence needs must have timely access to market information: the state of supply and demand, quantities and qualities demanded, current competitive prices, seasonal and unanticipated fluctuations in demand.

The ICSU-CASAFA report describes cultivable land as a fast-shrinking resource, and presses the need to raise food crop yields and farm productivity throughout Asia, Africa and Latin America. The report specifies the following urgencies:

a. to protect and enhance the natural resource base: farm land, surface and ground water, genetic diversity;

a. to reclaim and restore fertility in land degraded through misuse;

b. to promote systematic integration of production with post-production activities and to stimulate establishment and maintenance of rural agro-industries;

c. to devote more research to land capacity classification, soil and water management, to ecologically conservative land use.

The report illustrates the diversity, lack of consistency and many contradictions among definitions of, and prescriptions for, sustainable agriculture. It states: "It is unhelpful to propose any particular production system as sustainable in all ecologies, locations and societies. No specific system is suitable and sustainable for all circumstances ... Donor agencies should not seek to impose on poor farmers in developing nations systems designed to remedy wasteful practices in more affluent agricultural economies." The report contends that donors assign greater emphasis to agricultural production technologies than to post-production systems; to policy, social and economic considerations. The report urges far greater support for post-production systems; for technologies, logistics and economics of preservation, processing, transport, distribution and utilisation.

Other commentaries and recommendations

A publication commissioned for the 1995 World Food Summit cites various proposed prescriptions and purposes for sustainable agriculture [Hulse 1995]:

a. to generate agricultural products adequate and acceptable in quantity, variety and quality;

b. to maintain environments favourable to humans and other organisms;

c. to prevent pollution of surface and ground water; to protect wild life and animal rights;

d. to prevent despoilment and degradation of fertile land by erosion, urban spread and activities unfavourable to agriculture;

e. to establish and maintain rural infrastructures essential to agricultural production and marketing;

f. to protect natural ecosystems and ensure long-term conservation over short-term exploitation;

g. to promote nutrient recycling and ensure a balance between immediate use and long-term stability.

Some observers, mainly from developing nations, insist that 'sustainable' is not synonymous with 'low-input' systems that reject all chemical inputs. Nutrients derived solely from animal manure and composted biological waste may be insufficient to maintain fertility in poor soils, to support increased crop production needed to feed expanding populations. Modern technologies must serve to complement not replace traditional practices [Jain 1983].

Based upon his extensive international experience, Harwood [1990] presents an impressive historical review of proposed sustainable agricultural systems. He suggests the word 'sustainable' implies a steady state to be realised only by defining long-term goals. He states there are almost as many opinions about sustainable production as there are authors, many opinions being based on too little hard data and little understanding of variability in resources. He suggests many recommendations are simple and devoted to single issues rather than total complex systems. He considers it futile to seek any widely applicable definition of sustainable agriculture and proposes the following as a workable description of 'sustainable': "Agriculture that evolves indefinitely toward greater human utility, more efficient resource use, an environmental balance favourable to both humans and other organisms." Harwood deplores the excessive use of resources and lack of concern for ecologies evident in intensive production systems practised by many American farmers.

Conservation and preservation

As a general guide to what is sustainable, the following are deserving of particular attention [Swaminathan 2001].

- *Soil health*: physical, chemical, microbiological properties and susceptibility to erosion.
- *Water quality*: water to be used for irrigation must be of low salt concentrations.
- *Plant health:* crops must be protected from insects, other pests, pathogens and competing weeds. Pests can be persistent year round in tropical climates.
- *Genetic homogeneity*: genetically homogeneous crop types cultivated over large areas are susceptible to pests and pathogens that eventually overcome a plant's natural resistances.
- *Abiotic stresses*: soil salinity, flooding, water logging, periodic drought, chemical contamination are serious impediments to crop cultivation.
- *Post-harvest management*: uniformity in maturity, appearance, post-harvest stability during storage and transportation are properties neglected by governments and many donor agencies despite their greater importance to urban consumers and processing industries.

All harvested crops, livestock and fisheries are in varying degrees perishable, critical conditions being active moisture content and ambient temperature. In perishable produce, deterioration starts soon after harvest and in tropical climates progresses rapidly if preservation processes are not applied. As urban communities expand, as spatial and temporal distances between rural producers and urban consumers increase, the probability of post-production degradation increases. Primary processing close to where perishable crops are cultivated serves to reduce avoidable waste and to provide off-farm employment for rural people.

Swaminathan refers to the potentially adverse consequences of climate change, the need for genetic selections that reduce dependence on chemical inputs, research to devise remediation methods to reduce soil and water pollution, for genetic modifications to transfer such useful traits as tolerance to salinity from mangrove plants to cultivated food crops. Tolerance to soil salinity would extend Asian land areas suitable for crop cultivation.

Agricultural resources

Resources essential to agriculture may be classified as

Internal: Resources that exist and are renewable within a farm's confines;

External: Resources purchased, contracted for, and imported from outside the farm.

Resources can be sub-classed as a. natural, b. synthetic and manufactured, c. socio-economic, d. human [ICSU-CASAFA 1991].

Internal resources

Land classification and soil characteristics are critical determinants of practicable farming systems, systems influenced by a. land type and profile, whether flat, sloping or undulating, naturally wet or dry, with or without drainage; b. soil properties: structure, depth, density, chemical and organic composition, acidity, water retention and permeability. Efficient land management restricts erosion and structural degradation; maintains fertility and inhibits loss of plant nutrients. Prudent water management restricts pollution by agricultural chemicals and organic effluents, conserves surface, fossil and ground water, prevents over-drafting of rechargeable aquifers.

Internal flora include a. natural vegetation, some useful as forage or fuel; b. cultivated crops; c. soil and air-borne microorganisms, some that symbiotically stimulate plant growth, others that are harmful pathogens.

Internal fauna comprise a. natural wild life: i. productive organisms: insect pollinators, predators and parasitoids that attack crop pests, earthworms that improve soil properties; and ii. destructive organisms: arthropods and rodent pests, some as vectors for pathogens; b. livestock husbanded for food, fibre and on-farm work.

External resources

Whenever the load on carrying capacity exceeds the internal resources, external supplementary resources become necessary. External resources may be purchased outright or obtained under contact. External resources can be employed to modify land profiles, construct water catchments, plant trees, and provide machines and equipment for irrigation and other functions. External power from electricity or fossil fuels may be needed to drive machines; agricultural chemicals derived from fossil fuels provide fertilisers, costs being influenced by the world price of oil. Use of agricultural chemicals has persistently increased; excessive use can be hazardous to human health, to other organisms and to environmental stability. Bio-fertilisers, bio-pesticides and biological control of pests reduce dependence on agricultural chemicals [ICSU CASAFA 1991].

Agricultural land and soil

Soil is a living system: its structure, composition and biological diversity must be understood, protected and prudently managed if productive agriculture is to be sustainable [Brown and Wolf 1984]. The proportion of available arable land already cultivated varies from about 85 per cent across Asia to 22 per cent in sub-Saharan Africa. However, in Africa, land classed as arable includes large areas of tropical forest, extensive humid ecologies heavily infested with tsetse fly, the vector for the pathogens that cause sleeping sickness in humans and anaemic debilitations in cattle. Tropical forest soils are fragile and rapidly eroded when deprived of tree cover. Effects of soil degradation include debasement of physical structure; erosion by wind and water; depletion of organic matter and natural nutrients; toxicity from salinity, alkalinity, acidity and chemical contaminants. Much soil degradation results from avoidable, human carelessness. [ICSU-CASAFA 1991].

Of the world's total ice-free land, an area of roughly 13.4 billion hectares, the potential arable land is barely 3.2 bn ha [24 per cent of the total] of which about 60 per cent is of low productivity, susceptible to many stresses that inflict degradation. It is estimated every year some 20 million ha become uneconomic for farming, and an additional 6m ha are degraded beyond practical reclamation [IBSRAM 1991]. An international assessment of soil degradation reports that more than 1.9 bn ha of world soils have suffered some form of degradation, ranging from light to extreme, most resulting from mismanagement, deforestation, over-grazing, over-exploitation and industrial chemical pollution. Of the Planet's degraded soils, 38 per cent are in Asia, 25 per cent in Africa, 15 per cent in Central and South America, 11 per cent in Europe (including eastern Europe) and 5 per cent in North America [ISRIC 1991]. More than 110m ha of Indian soils are significantly degraded: 81 per cent by wind and water erosion, 13 per cent from loss of fertility, 4 per cent from salinity, 3 per cent from water logging. The area degraded represents about 30 per cent of India's arable land [Swaminathan 1992]. Of Canada's territorial area of 10 million square km, less than one per cent is high quality arable land. It is this land, best for productive agriculture, located in Canada's most benign climatic zones, that is most favoured for urban and industrial development. The World Resources Institute estimates more than 50 per cent of the world's arable land has been severely degraded [WRI 2004].

Agricultural systems must be sufficiently sustainable to provide for the needs of inexorably expanding populations, with greatest growth in urban communities while rural producers proportionately decline. However defined, by purpose or practice, sustainable agriculture is wholly dependent on fertile arable land and safe water, both disastrously degraded, wasted and misused. Arable land and unpolluted water are fast becoming the main determinants of future sustainable agriculture.

Variations on the sustainable theme

Much of the literature on sustainable agriculture originates in affluent industrialised nations and relates to intensive highly mechanised production systems that utilise substantial quantities of synthetic chemical inputs. Observers critical of subsidised intensive systems recognise in their nations less urgency to raise agricultural produce than to restrict disruption and degradation of natural ecosystems and resources. Though 'sustainable agriculture' is widely favoured, the following are other designations with reputedly similar objectives that have emerged in recent years.

Alternative agriculture

Literally interpreted, alternative agriculture describes any production system that differs from what is commonly practised. In some instances it is used to imply ecologically benign systems less dependent on agro-chemicals, other industrial inputs, more reliant on nutrient cycling, biological nitrogen fixation, propagation of natural enemies of crop pests and pathogens, conservation of soil and water quality, of biodiversity, protection of valuable symbiotic relations and biological interactions among crops, pests and pest predators.

Biodynamic

These favour systems that use compost and humus to enhance soil structure and fertility.

Ecological, Eco-biological, Socio-ecological

These are stated to be based on agricultural ecology harmonised with human ecology, to be protective of environments, to discourage use of chemical fertilisers, herbicides and pesticides.

Low-input

Advocates propose least possible use of agro-chemicals, fossil fuel energy, mechanisation, commercial certified seed and pedigree livestock. They recommend on-farm breeding and seed multiplication, composting of farm waste and crop residues, animal and green manure. Interpreted literally, low-input systems would revert to primitive labour intensive closed farming systems.

Low-input sustainable [LISA]

Proponents assert that productivity is raised by more efficient use of certified seed, energy, mechanisation and other off-farm inputs; by soil and water conservation, conservative tillage, multiple cropping mixed with livestock, systematic recycling of manure and crop residues.

Others designated 'Regenerative' and 'Permaculture' appear similar to 'Biodynamic' and 'Organic Agriculture'.

Organic farming

Strictly speaking, all agricultural production is 'organic' in that every product of agriculture is composed of organic substances. Among the earliest proponents of organic farming were two Britons, Sir Albert Howard and Lady Eve Balfour. Before World War II Howard advocated adoption of production systems that used no synthetic fertilisers or pesticides. For more than 30 years, on her research farm, Lady Balfour conducted experiments in which conventional British farming systems were compared with 'organic' systems free from agro-chemicals. Rodale, the American high priest of organic farming, may have received his inspiration from the published work of Howard and Lady Balfour [Rodale 1983].

Food products from organic farming represent a relatively small but slowly growing proportion of foods purchased in Europe and North America. Across the EU, the agricultural area devoted to organic farming has increased from 0.5 per cent in 1993 to *ca* 3 per cent in the year 2000. The value of organic foods sold in the United States is estimated at $10 billion, across the EU at 8 bn Euro. The total value of retailed processed foods in the United States is over $500 billion, in the EU more than 350 billion Euro.

The *Codex Alimentarius* guidelines for organic farming specify the following objectives:

- to enhance biological diversity;
- to increase biological activity and maintain fertility in soils;

- to restore soil nutrients by recycling organic wastes;
- to rely on local renewable rather than imported manufactured resources;
- to promote healthy use of soil, water and air;
- to minimise pollution from agricultural practices.

Despite the interest of environmentally conscious consumers, rational development of organic farming has been hampered by confusion among concepts and definitions, comparable to diverse interpretations about 'sustainable agriculture'. A comprehensive review and state of progress among EU countries has been published [ECDGA 2000]. Another European publication estimates sales of organically produced foods among EU members at about 1 per cent of total food sales. It also comments on the higher prices for organic foods charged in supermarkets. In France price differentials between organic and non-organic produce range from 10 to 40 per cent for dairy products, close to 50 per cent for fresh fruits and vegetables [EEC 2002]. In an 89-page document the European Council presents regulations that specify permitted production, labelling and inspection practices for organic foods produced and sold in EU markets [EEC 1991].

The general principles of organic farming are unquestionably admirable: harmonisation of agriculture with nature, protection of ecologies and environments, conservation of soil fertility and water resources, recycling of wastes, humane husbandry of livestock. At present organic foods are purchased by a small proportion of dedicated consumers who can afford the higher prices. Prices may gradually fall if and when demand and production increase. Whether or not strict adherence to EU regulations, total prohibition of all synthetic inputs and transgenic crops can ensure sustainable food security for all the poorest nations and communities is open to question.

Chemical fertilisers

While those who claim to practise or encourage organic farming vary in specific recommended practices, most urge systems more environmentally benign, relatively free from synthetic pesticides, growth hormones and antibiotics. Whether chemical fertilisers should be permitted is not consistent among organic farming advocates. Fertilisers provide nitrogen, phosphorus, potassium and other elements to soils naturally deficient, or to replace soil nutrients depleted by growing crops. Chemical fertilisation began with importation of guano from sea birds in Peru and nitrates from deposits in Chile.

In addition to the main constituents – carbon, hydrogen and oxygen – 13 elements are known as essential plant nutrients largely taken up through root systems. Most essential nutrients had their origins in the earth's crust from which soils were formed by weathering when nutrients were released, dissolved in soil moisture, then absorbed by the root systems of wild and cultivated plants. The chemical composition of the earth's crust is highly diverse across the Planet; soils in different locations vary significantly in plant nutrient content (ICSU-CASAFA 1991).

Chemical fertilisers are compounded to provide nutrients in which soils are naturally deficient or depleted by cultivated crops. Nitrogen, essential for all plant growth, originates in the 75,000 tonnes of the gas which float above every hectare of the earth's surface. Elemental atmospheric nitrogen cannot be taken up by plants directly but, by complex biochemical reactions and various organisms, must be converted to ammonia or soluble nitrates. Microorganisms known as Rhizobia live in root nodules of leguminous plants where they convert atmospheric nitrogen into ammonia to fertilise the host plant. In industrial processes, elemental nitrogen is 'fixed' to produce chemical fertilisers in a form that release nitrogen from soil as and when needed by cultivated crops.

Substantial energy is absorbed in conversions of nitrogen into ammonia. In a microbial conversion, energy liberated from fourteen moles of adenine triphosphate [ATP] is required to produce one mole of ammonia. In a Haber-Bosch industrial reactor, pressures of *ca* 20 million Newtons/m^2 at 500°C energise the combination of nitrogen with hydrogen. Some nitrogen-fixing bacteria obtain their energy from organic materials in soils; in legume root nodules Rhizobia derive their energy from the host plant in return for which they supply the plant with accessible nitrogen. As Rhizobia fix nitrogen in legume root nodules, therefore, some organic farmers rotate or intercrop legumes with cereals that lack the physiological and biochemical mechanisms to fix nitrogen. There is hope that biochemical capability to fix atmospheric nitrogen may eventually be transgenically implanted into cereal crops.

Other sources of nitrogen

Other nitrogen sources include animal manure; bio-fertilisers such as leaf mulch; compost from crop residues, in some instances assisted by earthworms as agents of conversion; green manure from leguminous plants and trees; such aquatic organisms as azola and blue-green algae. Manure, compost, mulch and other bio-fertilisers provide organic and fibrous materials beneficial to soil structure. Nutrient concentrations vary and rates of release from bio-fertilisers are not easily controlled, being influenced by soil temperature and moisture content. Release of nitrogen is generally slower from animal manure and composted residues high in lignin than from chemical fertilisers.

It is difficult to collect, store, distribute and utilise animal manure without significant nitrogen loss resulting from liquid run-off, leaching by rain, and volatilisation of ammonia. Manure stored in the open air in humid tropics with high rainfall can lose 70 per cent of nitrogen in 24 hours. Manure from cattle driven by pastoralists in semi-arid regions quickly dries and is blown away. Where wood is scarce and kerosene is expensive, animal manure may provide the cheapest source of household fuel. In contrast, chemical fertilisers are stable until used, designed and compounded to release nutrients as specifically needed in desirable composition and concentration.

Nevertheless, excessive application of chemical fertilisers may cause soil degradation, toxicity to earthworms and other useful soil organisms, and by leaching into ground and surface waters may cause insult to human health and harm to fish stocks. In northern European countries, where high cattle stocking densities are common, serious nitrification of ground water results from manure accumulations that freeze during the winter and release high concentrations of water-soluble ammonia during the Spring thaw. Farmyard manure is a source of various coliform microorganisms some of which are pathogenic. An outbreak of listeriosis in Canada was traced to raw salad vegetables fertilised with farmyard manure.

Contentious issues

Closed farming systems with crops entirely sustained by recycled nutrients from within the farm perimeter are appealing and deserving of encouragement. FAO estimates that agricultural production must grow by at least 3 per cent annually over the foreseeable future if all the world's people are to be adequately fed. It is doubtful if closed farming systems applied worldwide can come close to realising FAO's objective.

A defect frequently encountered in the 'sustainable' literature is failure to define specifically what or who is to be sustained. Is 'sustainable agriculture' to be interpreted as maintaining a constant level of agricultural productivity, or does it entail production consistently sufficient to sustain an expanding, diversifying population? As shown in Table 10, during the 1990s across the world, and in several regions, population grew faster than cereal grain production [WRI 2004].

Table 10
1990 to 2000 per cent change

	Cereal production	Average yield	Population	Fertiliser use
World	+9	+15	+16	[-3]
Asia	+19	+13	+19	+35
Africa	+18	+6	+31	[-5]
China	+10	+16	+15	+42
India	+21	+21	+25	+66

In at least ten African nations local cereal production decreased during the 1990s, a matter of serious concern since cereals provide the main source of food energy for the world's poorest people. The

International Food Policy Research Institute predicts the demand for cereals across all developing nations will rise by about 50 per cent between 1997 and 2020 [IFPRI 2002]. The IFPRI publication forecasts that between 1997 and 2020 average cereal yields in developing regions will drop from 1.9 to 1.2 per cent/year; demand for meat will increase by more than 90 per cent. Nutritionally, cereals are utilised more efficiently when eaten directly by humans than when fed to animals to produce meat for humans. Unless all livestock raised for meat are fed from natural pastures, food resources would be more conservatively utilised if all humans were vegetarians! IFPRI makes the gloomy prediction that by 2020, both China and India, the world's two largest nations that house almost one-third of all humankind, will become net importers of cereals [IFPRI 2002].

Sustainable agriculture: static or dynamic?

Agriculture cannot be considered sustainable if the quantity and quality of its products are inadequate to sustain the healthy survival of all who are dependent. Sustainable agriculture must be assessed by dynamic not static criteria. Simple maintenance of the *status quo ante* cannot ensure food security for expanding populations. From reliable assessments it appears that more than 5 million pre-school children throughout the world die every year from maladies caused or aggravated by malnutrition [IFPRI 2002]. Many causes of the melancholy state of, and forecasts for, future food and agriculture are offered:

a. a selfish indisposition of the affluent, both among and within nations, to be more conservative in their consumption habits, to share essential resources more equitably;

b. politicians who, to retain power and influence, invest more in armaments than to sustain more efficient systems of food and agricultural production and distribution;

c. sects more intent on inflicting their particular ideologies on the rest of mankind than in helping to alleviate poverty, social and economic inequities;

    d. self-proclaimed authorities who advocate a specific set of production principles or a production system sustainable for the whole world.

Sustainable agriculture in the United Kingdom

A report submitted to the British Minister for Environment, Food and Rural Affairs by a policy commission on the future of farming and food in Britain begins with a 'Vision' of desirable objectives and conditions for food and agriculture [PCFFF 2002].

    a. British farming and food sectors must be profitable, able to compete internationally, be good stewards of the environment, and provide a healthy, nutritious diet accessible to consumers in all income groups.

    b. Production, processing and marketing must be holistically integrated, with more effective inter-communication among them.

    c. The government is responsible for food safety, nutrition policy, animal welfare and environmental protection.

    d. Public policy must recognise the interdependence of town and country.

The Report then compares its idealistic 'Vision' with the 'Present Situation' and states:

    a. Farming and food industries [in Britain] are on a path that cannot be sustained over the long term.

    b. Because of subsidies and price supports imposed by the EU Common Agricultural Policy, British and EU consumers pay more for food than the rest of the world, inflicting particular hardship on poor consumers.

    c. The rural environment has been damaged by years of intensive agricultural production; while some farmers have adopted environmentally friendly practices, many cannot afford to do so.

    d. Though the 'Vision' emphasises more holistic integration, significant disconnection exists among farm producers, food processors and retailers; unsatisfactory inter-sector relations and communications are detrimental to overall sustainable efficiency.

e. British consumers are uneasy about the wholesomeness and safety of the foods they purchase.

f. many health problems among Britons result from people eating too much of the wrong kinds of foods.

g. farming has declined in importance as an economic activity, farmers' share of retail prices having fallen. The total food and agribusiness industry contributes about 8 per cent, farming contributes less than 1 per cent to GDP.

h. farming is a highly competitive activity; farmers must develop more sensitive communications with consumers.

i. over 90 per cent of foods are purchased from supermarkets, allowing them undue influence over farmers and consumers.

j. the government must accept responsibility to support more economically efficient, sustainable farming and food systems; the government's role is to facilitate essential change, not to finance stagnation and protection of the status quo.

All of the issues raised in the Vision and Present Situation are discussed in considerable detail throughout the report.

Sustainable Agriculture Initiative [SAI]

A few years ago the SAI was launched by the Unilever and Nestlé corporations, since when some 20 food processing and related agribusiness companies have become participants. Though there is sharing of general information, each participating company pursues its sustainable agriculture research and development independently with particular attention to raw materials essential to its brand products.

The quality of every processed food is critically dependent on the biochemical and biophysical properties of the raw materials from which it is processed. The properties of each raw material result from properties genetically inherited, degree of maturity at harvest, post-harvest conditions of handling, storage, transportation and the processing technologies applied in food factories. The SAI programme seeks to establish and maintain close cooperation between farm producers and processors: to improve and, where necessary, modify farming practices to ensure delivery to processors of raw materials of defined acceptable properties. SAI encourages farming practices conservative of critical resources, protective of land, water, the environment, safety and wholesomeness of all raw materials. Listed among SAI objectives are:

   a. to reduce risk of harmful residues such as pesticides, hormones and obnoxious contaminants;

   b. to ensure safe on-farm working conditions;

   c. to improve and expand farmers' knowledge and skills;

   d. to preserve soil fertility, minimise erosion, encourage minimum tillage;

   e. to promote farming systems that minimise waste and avoid pollution of surrounding environments;

   f. to reduce use of chemical pesticides by adoption of biological and biopesticide control;

   g. to use chemical fertilisers conservatively and effectively;

   h. to conserve and protect scarce water resources;

   i. to minimise energy consumed from non-renewable resources;

   j. to explore practical economic alternatives to fossil fuels;

   k. to protect biodiversity and wildlife habitats, the health and well-being of all farm livestock;

   l. to employ environmentally benign production systems;

   m. to raise farm productivity and profitability, to increase crop yields and reduce in-field and post-harvest loss;

   n. to devise efficient farming systems based on systematic analysis and simulation modelling of all existing and potential biological, technological, social and economic parameters.

The primary objective of the SAI is to ensure long-term availability of agricultural raw materials essential to their industrial processes, materials that provide more than 2/3 of all raw materials used by these industries. SAI participants' dedication to the above objectives is highly commendable, motivated by the need to restrict extensive degradation of arable land; recognising the competition for land from urban and industrial spread, and the desirability of protecting biodiversity, nature reserves and wild life habitats. SAI seeks to respond to the worldwide shortage of fresh water of which agriculture's consumption is close to 80 per cent of total world supply. SAI recognises growing competition from industries and urban communities, pollution of ground and surface waters by agricultural chemicals, municipal and industrial effluents, acid rain caused by industrial toxic emissions.

Guided by overall 'sustainable' objectives, SAI participants' R and D focuses primarily on agricultural crops and livestock of specific importance to their companies' processes. Unilever's concentration is on tea, palm oil, various vegetables and fisheries; Nestlé gives precedence to milk, coffee, sugar and cocoa; Danone to cereals and fodder crops. Environmental protection and resource conservation are also reflected in SAI participants' factory operations, where they pursue energy and water conservation, reduction and recycling of packaging and waste, elimination of refrigerants and other substances potentially harmful to atmospheric and environmental protection and conservation.

Though much maligned by 'anti-globalisation' activists, many transnational food and agribusiness corporations act more responsibly in conserving critical resources, protecting biodiversity, fragile ecologies and ecosystems, in improving agricultural practices and rural economies than can be said of some governments whose subsidies to farmers, damage to ecosystems and natural habitats in pursuit of oil and timber are destructive of ecologies and biodiversity. Governments and development agencies could benefit by following the SAI pursuit of economic, holistic integration of all logistical activities and processes that constitute a benign integrated food system that begins with sustainable on-farm production and ends with delivery of safe, nutritious foods to consumers.

International agricultural research: the CGIAR

Genesis and progress

A history of the foundation and early years of the Consultative Group on International Agricultural Research (CGIAR) was published by the World Bank [Baum 1986]. Other publications provide details of the growth and activities of the CGIAR and the family of International Agricultural Research Centres it supports [ICSU-CASAFA 1991, Hulse 1995, CGIAR 2002]. Annual Reports of the CGIAR and individual IARCs are available from the CGIAR Secretariat at the World Bank in Washington.

All of the CGIAR research centres protest their dedicated commitment to sustainable agricultural systems and practices [CGIAR 1995; 2002]. But it is not always clear whether the primary purpose is for each centre to be permanently sustained or to ensure that national agricultural organisations, the developing country farmers and the populations they serve are to be sustained.

The CGIAR had its genesis in agricultural research sponsored by the Rockefeller and Ford Foundations. In 1943, Rockefeller began a bean and cereal improvement programme in Mexico, which eventually brought forth semi-dwarf high-yielding wheat phenotypes and, in 1963, metamorphosed into the International Wheat and Maize Improvement Centre (CIMMYT). In 1960 the two Foundations established the International Rice Research Institute (IRRI) in the Philippines, and later two centres to study tropical agriculture, one in Nigeria, and a second in Colombia.

The CGIAR began in 1971 as an informal consortium of donor and development agencies with no legal constitution. From the start the World Bank has provided a supporting secretariat. Scientific assessments and recommendations come from the Technical Advisory Committee (TAC), a changing international group of experienced scientists with a secretariat housed at FAO in Rome.

In 1971 eleven members pledged $15 million to support the four existing IARCs. By 1976 membership had grown to 26 donors, who contributed $63 million to the four original and seven new IARCs. By 1992, CGIAR has expanded to a membership of 40 with a budget close to $300 million to maintain a family network of 16 IARCs and related programmes. The *CGIAR Report* for 2002 lists 62 donor members, including the three sponsors: World Bank, FAO and UNDP, and states there are now 16 IARCs that employ more than 8500 scientists and support staff. (A private communication indicates that in 2004 there were 15 centres and programmes, supported by over $400 million from 63 CGIAR members). The *2002 Report* refers to five priorities declared at the 2002 Johannesburg Conference: water, energy, health, agriculture, and biodiversity. Following a brief description of activities in each of the IARCs, the report describes Challenge Programmes in which several IARCS cooperate with emphasis on a. water and food, designed to increase productivity of water for food and livelihoods; b. bio-fortified crops for improved human nutrition: breeding crops with higher than normal contents of such micro-nutrients as vitamin A, iron and zinc. The report also describes a Global Conservation Trust, jointly sponsored by the CGIAR and FAO, to ensure diversity among the most important food crops through an endowment to provide funds sufficient to maintain crop genetic resources held in gene banks around the world. It mentions a report published by the International Peace Research Institute in Oslo that states that between 1989 and 1997 most of the major military and civil conflicts around the world occurred in regions heavily dependent on agriculture [CGIAR 2002].

A working group convened at FAO in 1989 by the CGIAR Technical Advisory Committee, with representatives from IARCs and several donor agencies, concluded that, "whereas agricultural production and farming systems have been well served by the CGIAR family, post-production systems, including breeding for functional properties essential to food processors, have been largely ignored."

CGIAR: Original purposes and progress

At the outset CGIAR and TAC agreed that the IARCs' research would be devoted to improvement of crops which provide most of the food for developing nations: crops such as rice, wheat, maize, sorghum, several grain legumes, cassava, potatoes and other root crops. Later IARCs were created for research on livestock health and husbandry, fisheries, forestry, agro-forestry, bananas, and a centre for collection and classification of plant genetic resources and the conservation of biodiversity. Each IARC responsible for one or more food crops maintains an extensive germplasm bank of broad genetic diversity. National agricultural research and development programmes can select and receive genotypes and phenotypes of crop species suited to their needs and ecologies. Each centre provides training in the disciplines and activities related to its research programme. The CGIAR and its IARCs have assembled the world's largest, most diverse collection of plant genetic resources, amounting to some 750,000 classified accessions.

The early advances at IRRI and CIMMYT, much publicised as the 'Green Revolution', resulted from Mendelian plant breeding which shifted synthesised biomass from vegetative plant parts into the edible seeds. Plant breeders also developed and propagated rice types that matured more rapidly than farmers' traditional varieties. Rice farmers were thus able to plant and harvest two or three rice crops on the same land every year. During the 1960s and 1970s, rice production across Asia more than doubled; India increased its wheat harvest from 11 million tonnes in 1961 to 70mt in 2002 [IEDRC 2003].

Much early criticism complained that those who benefited most from higher yields were large-scale not small-scale producers with limited resources. The world's history of technological progress illustrates that innovative and novel technologies are first adapted and adopted by individuals and industries with access to resources sufficient to tolerate the inevitable risk of investing in unfamiliar practices. Eventually, the higher yielding crops and related production technologies were accepted by a majority of Asian rice farmers [Chandler 1982; Baum 1986; ICSU-CASAFA 1991; CGIAR 1995; Swaminathan 1999].

Farming systems research

During the 1970s IRRI scientists laid the foundation for modern concepts of farming systems research, in a programme that began modestly with a few Filipino farmers but gradually spread to include many hundreds of rice farmers in 12 Asian nations. Western agricultural research traditionally began and continued on an experimental farm until promising results were eventually transmitted to farmers via extension services. In contrast, the IRRI farming systems research methods begin by assessing opportunities for, constraints and risks to increasing production by adaptation of new crop types and production technologies. The methodology starts with an analytical assessment of the biological, technological, physical, financial, economic and ecological resources existent among different farming communities; how and with what resources could more productive technologies be introduced with tolerable risk. The IRRI farming systems methodology has contributed significantly to increased crop production while conserving farmers' ecological and economic resources [IDRC 1981; Chandler 1982; Zandstra *et al* 1981] (*See* Case Study in Chapter 11).

Estimated benefits

Evenson and Gollin [2003] estimate results from what are variously designated 'high-yielding varieties' [HYV] and 'modern varieties' [MV] of food crops grown in developing countries. They compare changes in yield and total production during the 'Early Green Revolution (1961–80)' and the 'Late Green Revolution (1981–2000)' and the calculated influence of MVs mostly developed in CGIAR/IARCs. From 1961 to 1980 MVs accounted for 21 per cent of increased yield and 17 per cent of production increase; from 1981 to 2000, for 50 per cent of yield growth and 40 per cent of production increase. Most notable increases attributable to MVs were in Asia, the Middle East and Latin America. Little benefit from MVs was realised in sub-Saharan Africa where most increases resulted from expanding land under cultivation. Using a mathematical simulation model, the authors estimate that, without MVs in the period 1960 to 2000, food crop yields in developing countries would have fallen. The authors also claim that without MVs food prices would now be significantly higher throughout the world [Evenson & Gollin 2003].

Subcontracting to national programmes

An initiative that started in the International Centre for Research on Potatoes (CIP), and later pursued by IRRI, subcontracts programmes and projects to national agricultural research institutions in developing countries. The IARCs have trained thousands of scientists from poorer nations many of whom return to their native lands without funds and resources adequate to apply the knowledge and skills they have acquired. Subcontracting IARC programmes and projects to national research bodies serves a. to reduce operational costs, as developing country scientists cost less than their international counterparts; b. to increase resources and strengthen capabilities in national programmes. For agriculture to satisfy the needs of Africans, Asians and Latin Americans, resources and responsibilities should be systematically transferred from CGIAR/IARCs to national programmes and institutions.

Genetic modifications

Genetic modifications and pest control

All crops cultivated and wild are susceptible to attack by pests that infest, infect, predate or are parasitic. Most damaging are weeds and other competitive plants, arthropods (insects, mites, and spiders), pathogenic bacteria, fungi and viruses. Pests and predators cause insult and injury to vegetative and reproductive organs; pathogens inflict debilitating diseases; competitive and parasitic weeds deprive crops of essential nutrients; birds, rodents and insects wholly or partially destroy crops in the field and/or after harvest. Over countless millennia wild plants have evolved mechanisms and synthesised substances ('natural pesticides') by which to protect themselves from their enemies. Biophysical resistance mechanisms include tissues too tough or fibrous to permit penetration. Defensive biochemical substances are synthesised by growing plants: some toxic, some repellent to particular pests.

Cultivated plants and their pest enemies exist in dynamic relations. Pests mutate to overcome plants' natural resistance mechanisms and to tolerate chemical pesticides. To maintain resistance, plants gradually mutate to modify their effective defensive mechanisms. Being slower than insect and microbial mutations, it is among wild plants that stable evolved resistances are most evident. Cultivated plants can be protected by genetic modifications; by cultural practices such as intercropping where non-susceptible crops act as barriers to pest transmissions between same species; by biological control: propagation and release of organisms discovered to be enemies of particular pests; by chemical pesticides or bio-pesticides repellent or toxic to specific or a broad spectrum of pests. Protection by classical Mendelian breeding involves hybridisation of a susceptible genotype with a type resistant to particular pests. Modern transgenic techniques transfer and stabilise resistant traits from wild, distantly or unrelated organisms into cultivated species.

During the 1980s increases in food grain yields by classical Mendelian breeding were reaching practical limits; harvest yields were levelling off. In cooperation with the International Rice Research Institute, the Rockefeller Foundation developed and supported an international programme to devise techniques by which desirable genes could be transferred between distantly related or unrelated organisms. The Foundation acquired the active cooperation of an international group of experienced geneticists who worked to discover and devise innovative means of transferring genes between sexually incompatible species. The first objective was to enhance inheritable resistances in cultivated rice.

Of the 24 identified species of the *Oryza* genus, only two – *O.sativa* and *O.glabberima* – are cultivated as food crops. Over countless aeons the 22 wild *Oryza* species have evolved genetically inherited resistances to pests and pathogens to which cultivated species are susceptible. The objective of the RF-IRRI programme was to transfer genes that encoded for particular resistances from wild to cultivated *Oryza* species and thereby reduce rice farmers' dependence on chemical pesticides. Recognising that many persuasive activists suffer from extreme chemophobia, believing that any substance of biological origin is inherently less obnoxious and potentially harmful to human health than chemical synthetics, the RF/IRRI programme's objective of reducing reliance on chemical pesticides had logical reason.

Unfortunately, certain transnational agrochemical companies adapted the novel transgenic techniques to increase cultivated food crops' tolerance to chemical herbicides, thus permitting affluent farmers to apply higher than normal doses of herbicide to control competitive weeds. These commercial transgenic modifications were thus designed not to reduce but to increase demand for chemical pesticides.

Two international committees composed of experienced biological scientists pointed out that when a particular resistance is transgenically introduced into a cultivated plant, the plant's biochemistry is changed: it synthesises substances novel and foreign to its constitution, substances toxic or repellent to one or more pests. The target crop of the RF/IRRI programme was cultivated rice, the cereal that provides over 70 per cent of food energy for Asians. Much of the rice grown by small-scale Asian farmers is grown, harvested, distributed and eaten without scrutiny by a food regulatory agency.

Some years earlier, potato breeders in Britain and the United States had crossed standard cultivated varieties with Peruvian wild species to transfer resistance to certain pernicious pests. Fortunately, in both instances by happenstance, before the new hybrids were released to farmers and consumers, visiting scientists discovered that the resistances came from excessive and unsafe concentrations of a neurotoxic alkaloid. The new resistant types were immediately withdrawn and destroyed. With this experience in mind the two committees and a working group convened by the Canadian Food and Drug Directorate recommended to the rice breeders that, before release of any novel transgenic hybrids, biochemical analysts be engaged to determine the composition of the substances synthesised to provide pest resistance, to discover whether they were deposited in the edible seeds, or in vegetative plant parts. If in the edible seeds, toxicological and nutritional studies were essential to determine if concentrations present could result in insult or injury to the health of consumers. Had this advice not been totally ignored, subsequent opposition to GM crops might have been less clamorous, persuasive and restrictive of benefits that prudent genetic modifications could bring about [ICSU-CASAFA 1991, H & W Canada 1989, Hulse 1995].

Adoption of GM crops

A recent report by Clive James, a Canadian plant pathologist, states that in 2005, 54 million hectares (Mha) were planted to GM soybeans, 21.0 Mha to GM maize, 9.8 Mha to GM cotton and 4.6 Mha to GM rape-canola. GM crops are cultivated by roughly 8.5 million famers in 21 countries. The largest GM cultivators, in sequence of quantities produced, are the United States, Argentina, Brazil, Canada and China.

Disturbingly, James reports that 71 per cent of 90 Mha under GM crops are planted to herbicide-resistant genotypes of soybeans, maize, rape-canola and cotton. Though the original intention of the IRRI-Rockefeller GM programme was to reduce dependence on chemical pesticides, in fact the principal beneficiaries are herbicide manufacturers who also produce the herbicide-resistant seeds. The main active ingredient in the most extensively used herbicide is the isopropyl amine salt of glyphosphate. The potential toxicities of the active and other ingredients are still under study. In North America, the permitted residues of pesticides are scientifically prescribed and systematically monitored, but the same cannot be said of many poorer nations.

Assessing the safety of genetically modified crops

Persuasive publicity opposed to all genetic modifications, particularly among Europeans, whose survival is not threatened by food shortages, could prove detrimental to future agricultural research and to realisation of food security among poorer nations. While admitting that it is easier to discover acute or chronic toxicity than to demonstrate that a food or process can be assured as absolutely safe for all people under all circumstances, because of stringent food regulations, people resident among affluent nations have access to the safest food supply the world has ever known. But what may be demonstrated as 'safe' by reliable scientific analysis may not coincide with what is perceived to be safe by consumers influenced by strident propagandists.

The opportunities for crop improvement by molecular breeding and transgenic techniques are deserving of responsible scientific exploration. During the 1990s, it is estimated that more than 600 different chemical pesticides, valued at roughly $22 billion per year, were sold around the world [Dekker 1991]. Among nations of the EU, North America and Oceania, agricultural use of crop pesticides is rigorously controlled, while residues after harvest limited by law are systematically monitored. Among poorer nations, pesticide use is less stringently controlled and many instances of pesticide intoxications suffered by farmers and consumers have been reported. Broad-spectrum pesticides destroy not only targeted species but also many harmless and useful organisms. To induce inheritable resistances and thereby reduce use of dangerous pesticides is rational and sensible. Several scientifically reliable protocols to ensure safety of GM crops and derived foods have been published [Doyle and Persley 1996; IFBC 1990; IFST 1996; IFT 2000; IFT 2001; WHO 1991].

Natural pesticides

It has been estimated that some 5,000 'natural pesticides' – substances synthesised by plants to protect themselves – are present in foods sold in the United States. Though many have been identified, it is claimed that too little attention has been given to possible chronic toxicity, carcinogenicity, mutagenicity (disruption of DNA), and teratogenicity (induced congenital malformations) of many 'natural pesticides' [Ames

1983; Ames & Gold 1990]. An objective of many food plant breeding programmes was to eliminate or significantly reduce the concentrations of identified 'natural pesticides' known to be toxic, such as solanin the neurotoxin found in wild potato species. Cultivation of food crops deprived of their natural protection, require larger applications of chemical pesticides.

GM crops: potential benefits

A comprehensive review of opportunities for control of insect pests by molecular biology and transgenic modification concludes that these techniques will lead to improved pest management, facilitate precise studies of insect populations, genetic mapping, insect and crop resistance mechanisms. In addition to creating induced resistance in cultivated crops, genetic modifications of insects offer future means of biological pest control [Evans 1993].

Other potential benefits that genetic modifications of crops could provide include:

a. Greater tolerance to abiotic stresses – Tolerance to salinity can be transferred from mangrove plants into rice and other food crops [Swaminathan 2001]. Rice tolerant to soil salinity could substantially extend the land area across Asia for rice cultivation.

b. Enhanced nutritional quality – More desirable proportions of essential amino acids in cereal and legume proteins; higher contents of micro-nutrients deficient in many poor nations' diets [Swaminathan 2000, 2001].

c. Functional properties to expand suitability for food and other industrial processing – Scientists at the Plant Biotechnology Institute in Saskatoon, Canada, have modified the biochemical and biophysical properties of cereal starches to diversify their industrial applicability [Davis *et al* 2003; Demek *et al* 2000].

These are but a few of the possibilities for agronomic, nutritional and technological modification of food crops that could increase productivity and diversify utility of food crops for the benefit of poorer nations where, without significant improvement to protect and expand land suitable for cultivation, to raise harvest yields, and to reduce losses to pests and other degradations and debilitations, food security for many future millions in Asia, Africa and Latin America cannot be assured.

Closing comment

Practices and purposes deemed essential for 'sustainable agriculture' differ markedly among practitioners and commentators. As Harwood so cogently observed, there are as many prescriptions as there are authors. Commentators who live in affluent societies should be wary of prescribing remedies for poor folk who exist in fragile ecologies. The need to conserve biodiversity, arable land and water resources is undeniable. But no known crop or livestock production system is sustainable in all places and for all people. The IRRI farming systems methodology of acquiring a comprehensive understanding of what exists before offering remedies for improvement is a worthy example for all to follow.

Notwithstanding the many potential benefits to farmers and to food and health security offered by scrupulously designed and developed GM organisms, there is justifiable concern that the principal beneficiaries of agricultural GM are the producers of herbicides and herbicide-resistant food crops. Instead of reducing chemical pesticide use, herbicide applications are increasing.

In addition, it has been reported that some seed companies are experimenting with Genetic Use Restriction Technology (GURT), also known as 'Terminator' biotechnology. The plants are genetically modified to produce a sterile seed, so that farmers cannot save some of their harvested seed for later planting, but every year must buy new seed from the supplier. This practice is condemned under the Convention on Biodiversity formulated at the Rio conference, but there is evidence that some avaricious companies and the politicians they support and influence have hopes of ignoring the Convention.

8

Sustainable Food Security

Diverse definitions

As with 'sustainable development' there is no precise, universally accepted definition of 'food security'. In broad principle a state of food security exists where all individuals, families and communities enjoy consistent access to foods that in quantity, quality and biochemical composition provide hygienic, nutritional adequacy. The Brundtland Panel states: "Food security requires secure ownership of, or access to, food resources and income earning activities, including reserves and assets to offset risks, to ease shocks and meet all contingencies". This may be interpreted as all people having access to stocks and flows of food and cash sufficient to satisfy their basic nutritional needs. Everyone in the world need not be a subsistence farmer, but those who do not produce foods in quantities sufficient to satisfy their families' needs, must possess the means to acquire foods nutritionally adequate, readily accessible and affordable [Brundtland 1987b; ICSU-CASAFA 1991].

The report of the 1992 FAO/WHO international conference on nutrition states: "Food security is defined in its most basic form as access by all people at all times to the food needed for a healthy life.

Food security has three dimensions: assurance of a safe, nutritionally adequate food supply at national and household levels; a reasonable degree of stability in the supply of food throughout each year and from one year to the next; every household having physical, social and economic access to food sufficient to meet its needs."

While there is agreement in broad principle on requirements for food security, concepts diverge significantly in semasiological and biological specifics, they change over time, and differ among specialists and agencies as to what constitutes a nutritionally adequate diet. Non-scientific and quasi-scientific publications frequently fail to make clear distinction among 'hunger', 'malnutrition', 'famine' and 'starvation'; between 'chronic' and 'transitory hunger'. The noun 'security' derives from the Latin *securus* which means 'safe', 'free from worry and care'. People blessed with food security are safe from the risk of famine, starvation, chronic hunger and malnutrition. 'Famine' describes extreme food scarcity among large populations. 'Starvation' is derived from an Old English word meaning to die a slow lingering death from cold, disease, insufficient food and/or water. 'Malnutrition' defines a significant insufficiency or imbalance in essential nutrient intake [Latin *nutrire* = 'to nurse, suckle or feed'], 'sub-nutrition' sometimes describes people who have too little food. 'Hunger' describes distressing discomfort caused by want of food. 'Transitory hunger' is experienced from time to time by active persons who enjoy healthy appetites. 'Chronic hunger' describes a painful and debilitating condition caused by long deprivation of food sufficient to sustain healthy growth and activity [Hulse 1995]. The 1960s Freedom From Hunger Campaign, the 1974 World Congress' hope of eradicating hunger by 1984, and the FAO Millennium objective of reducing incidence of hunger by 50 per cent between 1990 and 2015, all relate to alleviation of chronic hunger and malnutrition.

Roughly 18 per cent of the world's population (800 million, including 200m children), 25 per cent of South Asians (300 million people) and 35 per cent of sub-Saharan Africans (190 million) are classified by UN agencies as undernourished: their caloric intake is lower than is required for healthy activity. In 1980 30 per cent of all people in LICs were considered undernourished. By the year 2000 this proportion had fallen to 18 per cent [FAO 2000].

Food security, income and poverty

In simplest terms, food security depends upon a dynamic balance among disposable income, demand, supply and distribution. Production of food crops, livestock and fisheries must progressively increase to satisfy expanding and diversifying needs and demands of growing populations. Efficient post-production systems, to ensure safe and effective protection, preservation, transformation and distribution must be economically integrated with production systems. During the early 1960s, affluent OECD nations were home to 34 per cent of world population and controlled close to 90 per cent of global GNP. By 1990, the affluent minority had fallen to 16 per cent of the total but controlled more than 82 per cent of global GNP. Between 1960 and 1990 a new group of middle-income newly industrialised countries emerged, most in East Asia and Latin America, a change with considerable implications for the Planet's economy and future patterns of food demand. In 1990 Asians represented 52 per cent, in 2002 some 59 per cent of world population. Among Asian nations, middle-income communities are growing most rapidly, their diets diversifying with dilating demand for livestock products.

Food insecurity, chronic hunger and under-nutrition are both dismal consequences and clear indicators of extreme poverty. People with money, power and control over negotiable assets rarely suffer from chronic hunger. Government politicians and military generals who suffer from chronic hunger are not a common spectacle around the world. Alleviation of food insecurity requires greater opportunities for paid employment and more equitable access to critical resources and assets among and within nations.

Changing and conflicting concepts

A respected observer writes: "In the 1950s … it was assumed that adequate agricultural production would assure access to adequate food in the market and in the household. In the 1970s it became clear that production and availability alone do not lead to food security, since those who lack purchasing power (from permanent employment income) lack access to a balanced diet. Now it is evident that where availability and access are satisfactory, necessary biological absorption of food requires clean drinking water, environmental hygiene, primary health care and education" [Swaminathan 2001].

Based on these considerations, the M. S. Swaminathan Research Foundation and the UN World Food Programme have published a *Food Insecurity Atlas for Rural India* [Vepa and Bhavani 2001]. The atlas illustrates marked variability among Indian States in food availability, absorption of nutrients, vulnerability to chronic hunger and in sustainable agricultural production. The atlas describes such non-food factors as income earning opportunities, health care facilities, education, sanitation and environmental hygiene.

Due to climatic variability, years with too little or too much rain, damage and destruction by pests, parasites and pathogens, crop and livestock production inevitably change from year to year. To compensate for years of poor harvests governments and/or regional organisations should maintain reserve stocks to carry over surpluses from regions and seasons of abundance to those of scarcity. The cost of holding reserve stocks should be calculated as an insurance investment not an unwarranted expense.

Food security need not be synonymous with food self-sufficiency, a nation's ability to produce all food required by its population. For diverse reasons: climate, topography, soil and climatic conditions, many nations cannot be self-sufficient and must depend on food aid or imports.

Substantial reductions in food spoilage after harvest could be realised by establishing primary food preservation and protection facilities in rural areas where crops are cultivated and animals husbanded. Food saved from spoilage increases total food available and enhances food availability and security.

Public policy and food security

Estimates of 'Global food security', based on the quotient of total possible food production divided by world population, may be a diverting statistical exercise, but is of little practical value. Equitable access to the planet's resources, though highly desirable, never has been nor is ever likely to be a realisable objective. To state that over the next half-century world food production must exceed all that was produced since settled farming began 10,000 years ago is interesting but offers little practical guidance for long-term planning by national governments. Food security can only be assessed and assured for discrete nations, communities and families within communities, a process that is the responsibility of national and local governments. International agencies can help weak and inexperienced governments by defining criteria to be considered and by providing long-term development assistance.

During the 1930s, the League of Nations Committee on Nutrition reported: "In the past, movement towards better nutrition has been largely the result of unconscious, instinctive groping for a better, more abundant life. Now needed is planning for better nutrition policies based on (a) consumption: bringing essential food within reach of all people; (b) adequate supply: production and equitable distribution of all essential food sources" [League of Nations 1936–1937]. The report urges (a) recognition of nutrition policy as of primary national importance; (b) better, widespread education on human nutrition; (c) more equitable distribution of income: "it is the poorest people who are most nutritionally deprived".

A committee of the US National Research Council appears to agree: "An important cause of malnutrition is absence of policies and programmes that foster the best use of available food supplies ... Governments continually make decisions that affect nutritional status with little knowledge of nutritional consequences. Alleviation of hunger and malnutrition depend upon increasing the right kinds of food, reducing poverty and improving the stability of food supplies" [NRC/NAS 1977].

Governmental responsibilities

Publications that discuss food security are many, vast and voluminous. They suggest primary causes, and propose various courses of remedial action. The League of Nations reports are among many that insist that food security be of highest priority in all governmental policies. Reutlinger [1987] states that food insecurity results from inappropriate macro-economic policies, from local economic and political structures that restrict the ability of households and individuals to have access to foods sufficient to satisfy their needs. Food insecurity may be transitory, lasting for relatively short periods, or chronic and of long duration, a consequence of poor planning to ensure resources adequate to produce or purchase sufficient food. Transitory hunger frequently results from unstable food production systems, sharp rises in food prices, or depression of disposable family incomes. If extreme and extended, transitory hunger can exacerbate into chronic hunger and eventually into famine. Famines have ravaged even when food stocks were plentiful: the 1943 famine in Bengal was precipitated by the price of a limited food supply rising to levels unaffordable by poor people.

Sen [1987] relates food security to entitlements and ownerships; families and individuals must have adequate access and purchasing power. Food security depends not on estimates of national aggregate demand but on fair exchange in international trade of goods and services. The poor must inevitably rely on food aid provided by social programmes. Chronic hunger and malnutrition are less influenced by a nation's total food supply than the degree of equity with which available resources are distributed and made accessible. Food and agricultural policies to ensure food security for all citizens should be a high priority for every government. Malnourished people, too weak to work effectively, eventually impose a financial burden on the nation's economy [Sen 1987].

Strategic food stocks

Though some observers argue that food stocks of grain are of value only to alleviate temporary food insecurity, the prime purpose of strategic food stocks is to provide emergency reserves against regional, national, local and seasonal shortages, to stabilise food prices and counteract fluctuations in supply and availability. Reserve food stocks are customarily controlled by government agencies, authorised to maintain quantities sufficient for defined time periods. Reserve stocks must be replenished when they fall below a prescribed safe minimum. Reserve stocks, essential in regions susceptible to frequent drought, provide short-term relief but cannot be considered a panacea for chronic food insecurity.

Santorum and Gray [1993] list factors to be weighed in determining minimum reserve stocks. Stocks stored on farms are generally sufficient only to satisfy farmers' needs from one harvest to the next. Countries within a region that cannot individually afford to maintain national reserves can collectively contribute to and draw upon strategically dispersed food stocks. During the 1980s the Southern African Development Community (SADC) created and maintained a regional food security programme with grain stocks upon which member nations could draw in time of need.

Food aid

Where many are too poor to buy adequate food, where climatic change, civil strife and other adversities inflict serious disruptions on food production and distribution, food aid may be essential to survival and avoidance of famine. The UN World Food Programme [WFP] is the most efficient distributor of food aid, designed to provide relief or to support development. Food aid for relief alleviates acute stress among people whose food supplies are disrupted by war, civil unrest, enforced migration, severe climatic catastrophes, and/or massive crop failure. Food aid for development provides food as wages for economically productive work and to support basic social services. Food aid for development may support labour-intensive agricultural production, construction of roads, rural infrastructures, amenities and services essential to developing economies. Food aid helps to stabilise food prices, to provide for pregnant women, nursing mothers, school feeding and pre-school infants.

WFP administers food aid most efficiently. Misdirected, ill-conceived bilateral food aid can do more harm than good. One could cite a weary litany of bilateral food aid where indiscriminate dumping of donors' surplus crops and livestock products have depressed prices and discouraged agricultural production in recipient countries. Food aid administered by many bilateral agencies is motivated more by the donor's political and commercial interests than relieving acute or chronic distress among malnourished people.

WFP resources have been seriously overstretched, because of exceptional demands resulting from military, civil and inter-ethnic strife and conflict, mass migrations such as in Darfur, cataclysmic disruptions as produced by the Tsunami. As extreme demands for food aid will inevitably persist, more generous and sustained support for the World Food Programme should be a high priority for all donor agencies.

Nutritional adequacy

What constitutes nutritional adequacy has attracted scientific and unscientific interest for many centuries. Alchemists searched for but never discovered the *Elixir vitae*. In Britain during the mid-nineteenth century desirable dietary standards were described as: "the least costly foods to prevent starvation and disease during massive unemployment". The League of Nations standards were intended "to marry health and agriculture" [League of Nations 1936–1937]. Our most primitive ancestors, by trial and error, identified certain foods as pleasant and digestible, others as obnoxious, insulting or injurious to health, foods acceptable and foods unacceptable. Our early antecedents chose their foods from plants and animals indigenous to their ecologies and climates resulting in widely different food preferences among nations and communities.

Dietary standards and recommendations

Estimates of human nutrient requirements were first codified during the latter half of the nineteenth century. From the time of their establishment, FAO and WHO have sponsored many meetings, committees and working groups to study and recommend desirable nutrient intakes, each meeting re-estimating recommended levels based on new knowledge. At least 50 national, regional and international agencies recommend what they consider desirable intakes of essential nutrients. Recommendations are customarily cited as specific amounts of each essential nutrient to be ingested daily. Some authorities name them Recommended Daily Allowances [RDAs], others Recommended Daily or Dietary Intakes [RDIs], others Recommended Nutrient Intakes [RNIs]. The evolution of recommended dietary standards was reviewed by several authors [Beaton 1991; Hulse 1995; IUNS/CABI 1983]. The purpose of RDAs is to encourage healthy eating habits and to prevent diseases attributable to nutrient deficiencies. Earliest motivations for government dietary standards arose in times of severe stress and distress; to feed nations at war or when suffering economic depressions, widespread poverty and malnutrition.

Recommended Daily Allowances [RDAs] the preferred American designation, are intended to serve several purposes: (a) to assess adequacy of food intake among selected population groups; (b) for national food and agricultural policy formulation; (c) to balance food nutritional quality with human need; (d) to provide adequate food during emergencies, particularly for the poor and disadvantaged; (e) to prescribe and enact standards for nutrient labelling of packaged foods. RDAs were first designed to meet the needs of the majority, often embodying margins of safety to allow for variable needs among people of different ages, activities and conditions of health. It is now customary to prescribe specific RDAs for citizens of different sexes, ages and levels of activity.

RDAs were first calculated by estimating the average physiological requirements for each absorbed nutrient, a value then modified to compensate for incomplete absorption and utilisation, variability among persons, bioavailability among different food sources. RDAs are based on foods consumed in normal diets, not on pure substances or concentrates, and take account of nutrient variability within and among particular foods. At high levels of intake some nutrients, such as vitamin A, may be toxic so that upper limits of safe intake are prescribed. It is difficult to prescribe precise RDAs for all conditions of men, women and children and their variable activities. Experiments on human beings are expensive, time consuming and only small groups of volunteer subjects can be studied in confined controlled conditions. Using humans as experimental animals also raises ethical objections. RDAs are intended for people as they go about their normal daily activities but food intakes and nutrient balance are difficult to determine where subjects move freely and are not confined in a metabolic research facility. Data based on *ex post facto* recall by subjects going about their normal lives are at best approximate.

Despite a vast accumulation of nutritional data, with more sensitive analytical tools available, nutrition remains a relatively imprecise and speculative science. Concepts change as more evidence comes to light. Thirty years ago it was believed military men working in Arctic climates needed more food than those working in temperate zones so Canadian troops in the Arctic north were given packaged rations that provided 5,000 kg Cals/day. It was later realised that with insulated clothing and the need to move slowly to avoid heavy perspiration that could freeze, caloric requirements were no higher in cold than in temperate climates. Since caloric intake must balance energy expenditure, people who perform heavy work or strenuous exercise need higher caloric intakes than sedentary persons.

People who change from energy-intensive activities to a sedentary life-style tend to eat more than they need. Overeating with resultant obesity is disturbingly evident among the richer nations and affluent persons in less developed economies. It is reported that between 25 and 30 per cent of Europeans and North Americans are excessively obese, at risk of cardiovascular diseases, diabetes and other related disabilities. The obese among the affluent reflect Nerissa's comment in Shakespeare's *Merchant of Venice*: "They are as sick who surfeit with too much as those who starve with nothing."

Nutritional fashions

Responsible nutritional scientists seek to encourage rational eating habits based on the best available knowledge. The alchemist's pursuit of the *Elixir vitae* is far from moribund among publicists who proclaim all manner of diets guaranteed to ensure good health and longevity [Hulse 1995]. The penchant for nutritional panaceas has been evident for the past 50 years. During the 1950s vitamins were in favour, in the 1960s proteins and amino acids were in fashion, from the 1970s cholesterol and cardiovascular dysfunctions obsessed public attention, next came the health benefits promised by dietary fibre. The present passion is for nutriceuticals, proclaimed in principle some 4,000 years ago by the Chinese, and for 'functional foods' (which beg the question: what are non-functional foods?), neither of which are specifically defined.

Variability in Composition

The nutrient content of every food crop varies among genotypes and conditions of cultivation. Among the world's wheat collection on a dry weight basis, protein contents range from 6 to 23 per cent, lysine (as per cent of protein) from 2 per cent to 4.3 per cent [Hulse and Laing 1974]. A review of over 1,300 sorghum analyses reported protein contents (N × 5.7 DWB) to range from 7.1 to 14.2 per cent, lysine (mg/g nitrogen) from 71 to 212, iron (mg/100g) from 0.9 to 20.0, niacin [mg/100g] from 2.9 to 6.4. Significant variability was reported among samples of pearl millet (*Pennisetum typhoides*), chickpea (*Cicer arietinum*), faba bean (*Vicia faba*), and lentils (*Lens esculenta*) [Hulse 1991a, 1994, 1995; Hulse and Laing 1974; Hulse *et al* 1980].

Accurate chemical and biochemical analyses are difficult and demand professional competence and experience. Inexperienced technicians are unaware of many sources of error among standard analytical methods. Some 16 sources of error were identified in the classical Kjeldahl method of determining protein nitrogen [Williams 1974].

Changing recommendations

Since the nineteenth century, recommended daily intakes of all known nutrients have changed among and within responsible agencies. In 1985 a WHO committee recommended a daily protein intake of 0.6g per kg of body weight; a US NRC committee advocated 0.75 g per kg of body weight. Table 11 illustrates how US RDAs have changed since first promulgated in the early 1940s.

Table 11
US NRC Recommended Daily Allowances

	1945	1989
Protein (g/day)	100	59
Ascorbic acid (mg/day)	100	60
Iron (mg/day)	15	12

Table 12
UK Recommended Daily Nutrient Intake

	1950	1979	1993
Vitamin A (RE* mg)	1500	750	800
Ascorbic acid (mg)	20	30	60
Niacin (mg	12	15	18
Iron (mg)	12	10	14

* RE = retinol equivalent

A comparison of daily intakes recommended by 39 governments and three international agencies, updated in 1995, illustrates many inconsistencies. Table 13 lists intakes prescribed for young active males (M), for teen-age women (F). Disparities of similar magnitude were evident for all other ages and levels of activity.

Table 13
Recommended nutrient intakes (RNIs) from 41 countries and agencies

		Highest RNIs	Lowest RNIs
Energy (kcal)	M	3200	2500
	F	3000	2030
Protein (g)	M	105	37
	F	96	31
Vitamin A (mg RE)	M	1500	600
	F	1500	450
Vitamin D (mg)	M	10	2.5
	F	10	2.5
Ascorbic acid (mg)	M	95	30
	F	85	25
Iron (mg)	M	24	5
	F	35	12

Many more people would be classed as malnourished if the highest rather than the lowest or some level in between were the international standard. For international comparative purposes it is sensible to use the intakes prescribed by WHO and FAO. In light of these many differences and uncertainties, all RDAs, RDIs and RNIs should be regarded as indicative but not scientifically absolute. Disparities among nations may reflect different life styles and activities or various differences of opinion. A Canadian publication states: "The primary use of nutrient recommendations is to form a basis from which to advise the public on how to choose a healthy diet. It would be a mistake to conclude that those who fail to adhere to each recommendation are doomed to suffer associated chronic deficiency disease, or that by strict adherence consumers will avoid deficiency diseases" [H and W Canada 1990].

Recommended Daily Allowances (RDAs)

RDAs, RDIs and RNIs can be useful as general guides for cognisant officials responsible for formulating national health, food and agricultural policies. For most consumers, RDAs have little practical meaning. Consumers are better advised by illustrated guides to healthy eating based upon mixed diets of foods that are familiar, acceptable and seasonably available [H and W Canada 1992]. Human nutritional requirements, criteria used to arrive at recommended levels of intake, consequences of serious deficiencies, food intolerances and allergic responses are discussed in greater detail in other publications [Beaton 1991, IUNS/CABI 1983,Hulse 1995].

Urban growth and food security

Predictions of the size to which the world population will grow, and for how many persons the planet can provide sustainable food security 50 or 100 years from now, though intellectually diverting, are of little practical utility. One can estimate with reasonable certainty what could be produced with present resources. What may be feasible 50 or 100 years from now is indeterminable.

Of more pressing urgency are the consequences of rapidly changing demographics and rising disposable incomes among urban communities. More than half the people of Asia and Latin America live in large towns and cities, expanding in size and populations all over the world. Though some food can be raised in city gardens and horticultural allotments, urban communities rely on food safely transported from rural farms, inland and maritime fisheries.

As people migrate from rural to urban communities and disposable incomes rise, eating habits change. Growing demand for processed convenience foods is illustrated by the replacement of cereals, pulses and cassava, hand ground and cooked into some form of porridge, with processed foods such as bread, biscuits, cakes and pasta. Of critical consequence for natural resource conservation is the growing demand for livestock products, which across SE Asia is rising by over 20 per cent a year.

Few of the world's cities were planned and laid out in expectation of the volume and density of vehicular traffic destined to grow progressively worse in all cities on all continents. To alleviate traffic congestion, the City of London Authority has imposed heavy tolls on private vehicles that enter the city on working days. Too little investment is devoted to improving and maintaining rural infrastructures, to rural-to-urban and intra-urban transportation systems. The longer time taken to transport perishable food materials from rural producers to urban processors and markets, the greater is the probability of biological and economic loss.

Urban food security demands consistent access to a safe adequate supply of acceptable, affordable foods dependent on a. technologically efficient systems of preservation and processing, primary preservation being applied soon after harvest; b. logistically efficient and economic means of storage and transportation; c. reliable markets which retail foods that are safe, of acceptable quality, in unit quantities and prices that all segments of urban society can afford; d. national food and agricultural policies that provide and maintain the infrastructures, services, transportation facilities, social and economic conditions essential to efficient holistically integrated food systems.

Final observation

The Brundtland principle of providing adequately for present need without detriment to the needs of future generations has particular relevance for food security. The future need, globally and among specific communities, and future agricultural capacity cannot be predicted, it is therefore, incumbent on all peoples, their governments, industries and institutions, to ensure that resources critical and essential to food production, preservation and distribution be protected and used conservatively. Degradation of arable land suitable for crop production must be arrested, where possible degraded land must be reclaimed. Water, the most critical and scarce of all essential resources, must be protected from pollution and misuse, whenever feasible purified and recycled.

In expectation of unrelenting inter-ethnic conflict, massive migrations, increasing severity of damage by hurricanes and cyclones, together with severe droughts, all consequent upon climate change, the need and demand for food aid is more likely to grow than contract.

More generous and sustained support for the World Food Programme must be accepted as an obligation by all affluent nations and donor agencies.

Assurance of sustainable food security for expanding congested urban populations demands efficient integration of all activities and procedures between rural producers, food processors, urban markets and consumers. Urban food security in all nations calls for systematic planning by professionally qualified men and women of imaginative vision. For, as the *Book of Proverbs* so wisely observes: "Without vision the people will perish".

9

Industrial Biotechnologies

Biotechnologies: definition and evolution

Biotechnologies are defined as 'Processes that produce, preserve and/or transform biological materials of animal, vegetable, microbial or viral origin into products of industrial, commercial, economic, social and/or hygienic utility and value'. Biotechnologists and bioengineers are men and women qualified to design, develop, operate, maintain and control biotechnological processes.

The name 'biotechnology' first appeared in 1897 in Yorkshire, England, where a Bureau of Biotechnology provided consultant and analytical services to local fermentation industries. The early development of the principal biotechnology industries – food, drugs and textiles – display marked similarities, though their modes of development have differed over the centuries. All began using empirically derived artisanal labour-intensive technologies which, following different patterns, were gradually mechanised, human hands being replaced by mechanical devices and machines driven, successively, by animal power, water, wind, steam engines, fossil fuels and electricity.

For many centuries, technologies and industries that processed foods and drugs have been closely associated. Standards of quality and safety for foods and drugs are frequently administered by the same regulatory agency, the US Food and Drug Administration being a typical example.

Beginnings of biotechnologies

Sometime around 3,000 BCE the Chinese began acting on their belief in a close association of foods with medicines, both essential to good health, both at the time derived from plants and animal organs. The Chinese continue to believe that many ailments can be cured by appropriate diets; they were the first to add burnt sponge, an aquatic source of iodine, to people suffering from goitre. (The present fashionable addiction to 'nutriceuticals' is simply an extension of an ancient Chinese doctrine).

The earliest drugs were discovered by perceptive observation, trial and error, certain plants and their extracts found to alleviate pain and particular symptoms of disease. Some medicines still prescribed were discovered more than 4,000 years ago. Ephedrine extracted from a Chinese shrub *Ephedra sinica* was then and is now prescribed to relieve asthma and other allergenic respiratory dysfunctions. Between 4,000 and 5,000 years ago Egyptians prescribed remedies for more than 1,000 ailments. The Egyptian Ebers Papyrus [*ca* 1500 BCE] lists over 800 drugs compounded from such vegetable extracts as gentian, castor oil, vermifuges (anthelminthics to treat intestinal parasites) and henbane. The Papyrus prescribes treatments for rheumatism, schistosomiasis and diabetes.

A characteristic common to food, drug and textile industries is that empirically discovered and derived technologies long preceded any scientific understanding of the biochemical and biophysical properties of the raw materials and processed products. For many centuries food and textile industries progressed through mechanics, the gradual replacement of human hands by machines. The principles underlying industrial processes of food fractionation and transformation, the spinning and weaving of natural fibres, were discovered several thousand years ago. Progressive mechanisation, which at first led to unemployment and deprivation among rural artisans, has significantly economised on human effort and energy expenditure in factories, restaurants and homes.

Food preservation

The basic principles of food preservation that involve control of a. active water content, b. ambient atmosphere and temperature, c. pH, d. thermal inactivation of microbial and biochemical deterioration, were discovered long ago. Aboriginal, Asian and Mediterranean people used sun-drying to preserve milk, meat, fish, fruits and vegetables; sliced potatoes were freeze-dried by early Amerindians, from which the ice crystals sublimed in the dry air and low atmospheric pressures of the high Andes. Stone-age Britons dried moist grains over open fires to prevent sprouting. More than 4,000 years ago the Chinese preserved fish by osmotic dehydration with salt. Republican Romans reduced water activity in meat and fruits by adding salt or honey. About 5,000 years ago Middle Eastern farmers stored grains in filled earthenware amphora, hermetically sealed by impervious goat-skins, causing all stages of insect metamorphosis to be asphyxiated by respired carbon dioxide. Seneca describes how Romans preserved prawns and oysters in snow collected from the Apennines.

The modern frozen food industry developed after an American, Clarence Birdseye, observed how Aboriginal people in northern Canada preserved surplus whale, seal and reindeer meat in ice collected during the cold winter. Modern canning, bottling and boil-in-the bag were anticipated in Republican Rome by chopped spiced meats sealed and boiled in the cleaned womb from a sow or the body cavity of a squid. Preservation by fermentation and pickling was practised by Egyptians and other ancient Mediterranean people. Babylonians preserved all their milk by lactic fermentation. Fermented bread, beer and wine were produced on a factory scale during successive Pharaonic dynasties, while ethanol was distilled in China 3,000 years ago.

From hand labour to mechanisation

Textile technologies

The patterns and pace of mechanisation moved differently among the biotechnologies, textile industries progressing faster than food processing. Spinning and weaving of fibres from cotton, wool, flax and silk were practised 6,000 years ago in Egypt and Mesopotamia, the yarns being spun from a distaff, a cleft stick, onto a spindle. Spinning being exclusively an occupation of women, the word 'distaff' has long been associated with femininity. With relatively simple equipment, more than 4000 years ago Persians wove wool and cotton for clothing, luxurious silk fabrics, ornate woollen carpets and tapestries. Evolution from domestic spinning and weaving in rural homes to mechanised factory operations in Britain took many years. The compulsion to assemble workers and machines in factories arose from the need for cheap energy and to enforce more consistency in quality and uniformity. Until horse-drawn barges along canals carried cheap coal to energise steam engines, textile factories, in common with grain mills, were positioned on river banks driven by power transmitted from water wheels.

Textile technologies were mechanised by a series of inventions: John Kay's flying shuttle patented in 1703, various spinning machines invented in the late eighteenth century. From the Middle Ages rural women had spun wool and flax with the Saxon spinning wheel. Crompton's mule, invented in the 1760s, enabled one spinner to control 1,000 spindles, thus casting countless rural women and their families into unemployment.

None of the men who invented textile machines and none who contributed to invention and development of the nineteenth century steam engine were scientists or professional engineers. Thomas Newcomen was a plumber whose steam engine, designed to pump water out of coalmines depended on a piston driven by steam injected into a cylinder. By more accurate boring of cylinders that housed the piston and other devices, James Watt, an instrument maker, improved the efficiency of Newcomen's engine. One of Watt's early engines drove machines in a London flour mill, others gradually replaced water power in textile mills.

Ancient Egyptians dyed fabrics with plant pigments and the rich purple from shells of the aquatic murex snail discovered in the Mediterranean by the Pheonicians. During the Dark Ages the art of dyeing was lost among Europeans but was revived in the late twelfth century when Flemish dyers rediscovered plant pigments, their most spectacular being pigments from the roots and petals of madder (*Rubia tinctorum*) and the Asian lichen orchis (*Roccella tinctora*). From the mid-nineteenth into the twentieth century chemistry provided the greatest impetus to textile development, first through German synthetic aniline dyestuffs, later through synthetic fibres; viscose rayon regenerated from chemically modified cellulose was among the first.

Since the eighteenth century inventors and engineers have designed faster textile machines, with higher spindle speeds, direct winding onto looms and, most recently, fabric and garment design and manufacture controlled by computers. Over the past 50 years scientists have discovered the chemical composition of natural fibres, the changes that take place during processing, subsequent wear and tear, and from environmental stresses. Mathematicians have contributed to physicochemical analysis of the complex structures and changes involved in garments made up of millions of fibres largely held together by frictional forces. Plant and animal scientists breed for improved fibre quality in cotton and sheep wool, chemists synthesise new fibres and fabrics, computer specialists and mathematicians devise programmes to design fibres, fabrics and garments and to control the machines that produce them.

Textile biotechnologies have developed from simple domestic and artisanal crafts into scientifically controlled and highly mechanised integrated industries. But as with most industrial biotechnologies, progress to the present state took place over 7,000 years; innovations were devised by people who perceived opportunities for their inventive talents. Development proceeded spasmodically and episodically, knowledge sometimes passed on between generations, sometimes lost and then rediscovered. As disturbingly described by social historians and nineteenth century novelists, the pursuit of profit through mechanisation of textiles and other industries inflicted immense hardship and suffering on men and women displaced by machines. The horrors of the Industrial Revolution illustrate the persuasive need to assess systematically the social consequences of all development programmes.

Grain and oilseed milling

Milling of grain was the first food process to be mechanised, which is not surprising since grain fractionation by hand pounding and winnowing is the most tedious and time consuming of all food processes. Fractionation of grains by sieving and winnowing, extraction of olive and linseed oil by pressing was practised in Egypt and Mesopotamia more than 7,000 years ago. Commercial bakeries and breweries existed in Egypt and Mesopotamia 5,000 years before Eduard Buchner and Emil Fischer discovered the enzymatic conversion of carbohydrates into ethanol and carbon dioxide.

The earliest cereals to provide food for Mediterranean people were probably the seeds of wild emmer and einkorn, the outer hulls of which were tough, indigestible and so firmly attached that grains had to be parched over hot fires before being hand-pounded between a concave rock and a stone club, the earliest ancestor of the pestle and mortar. Some time around 8,000 BCE, naked wheat grains, free of outer hulls, the probable outcome of promiscuous transgenic hybridisation of emmer, einkorn and other grasses, were adopted and eventually cultivated [Hutchinson 1974; Simmonds 1976]. Primitive pestles and mortars were replaced by saddle stones, an early Egyptian device with which a kneeling slave propelled an upper stone with a lower smooth flat surface to crush grains spread over the concave surface of a lower stone slab. The shearing action to break open the grain was gradually improved by incising a herring-bone grooved pattern onto the working surfaces of the upper and lower stones. Later, attaching a wooden handle to move the upper stone backwards and forwards, and an eye-hole in the upper stone permitted a slave to feed grain continuously while moving the upper grinding stone. Grain milling thus became the first mechanised continuous food technological process, raw material fed and end product emerging in continuous flow.

Republican Romans adapted the Egyptian rotary quern in which an upper stone (the catillus) rotated over a lower stationary stone (the meta), grain being fed continuously via an eye-hole in the centre of the upper stone. Milled grain moved to the periphery by centrifugal force of rotation. Through a geared drive shaft, Roman querns were powered from water wheels. The first windmills were devised during the tenth century CE in what is now Iran and Iraq. Mills driven by water and wind power were dominant well into the nineteenth century; it is less than 100 years since grain mills powered by steam engines in North America were equal in number to mills driven by water wheels. In 1784 an early model of James Watts' steam engine was installed in a London wheat flour mill; the first mill driven by electric motors was constructed in 1887 in Laramie, Wyoming.

Though more precisely engineered than their ancient ancestors, modern grain mills equipped with incised break rolls are direct descendents of the Egyptian saddle stone and the Roman quern. Hand winnowing has its modern counterpart in an enclosed vibrating gravity table, combined with screens of various mesh size to fractionate particles of different dimensions. An ingenious software programme, devised in Bangalore, allows modern mills in India to be operated and controlled from a desk-top computer, while among poor rural Indian communities local grains are still ground by large wooden pestles and mortars, by saddle stones, and groundnut oil is extracted by animal-powered rotary querns.

The protein content of wheat, oilseed and grain legume flours can be enhanced by fine grinding and air-classification. In cereal, legume and oilseed grains starch granules, of varying size and effective mass, are embedded in a protein matrix. Fine grinding by dynamic shearing devices cause some starch granules to separate from the protein, which is shattered into fine fragments. By air classification, in which a mechanical centrifugal force is opposed by a predetermined centripetal air drag, the lighter protein fragments and the heavier starch granules move in opposite directions. The starch granules in grain legumes and oilseeds being larger than in cereals, larger protein shifts can be produced from the former than the latter.

Science applied to biotechnologies

As with stone tools, biotechnologies evolved through three overlapping periods. The ancient empirical period began at least 10,000 years ago and evolved until the early eighteenth century. From then the middle period progressed as greater scientific understanding of biologies and biotechnologies gradually emerged. The modern Genomic Era began in the early 1970s. Systematic scientific study came late to industrial biotechnologies. The early history of food science, which progressed spasmodically during the eighteenth and nineteenth centuries, consisted of a gradual revelation of the composition and properties of food materials. Not until the twentieth century were novel food and textile processes devised directly from scientific principles. In contrast, during the nineteenth century pharmaceutical industries developed from the application of chemical knowledge learned in the German dyestuffs industries. Historical reviews of biotechnological progress are in earlier publications [Hulse 1995, 1996, 1998, 2003].

Evolution of pharmaceuticals

Primitive people searched for natural panaceas and palliatives to cure their diseases and relieve their discomforts. Most early drugs were extracts of wild plants, some animal organs and a few available chemicals. Sumerian cuneiform tablets from the time of Hammurabi describe hepatic and febrile diseases, gonorrhoea, scabies, strokes and seizures. Babylonian drugs included extracts of hellbore, believed to cure psychological disorders, mandrake root and opium derived from the latex of certain poppies (*Papaver somniferum*).

From earliest times, sickness was regarded as punishment for behaviour offensive to the gods; curative medicine often being superstitiously associated with magic. During the fourth and fifth centuries BCE the Greek school of Hippocrates proposed over 300 remedies, most being plant extracts administered orally or via other body orifices. Aware of magical associations and that drugs in excessive doses could be fatal, the Greek word φαρμακον (pharmakon) can be translated as 'medicine', 'drug', 'poison' or 'magic potion'. The Hippocratic oath demands that in practising their diagnostic and curative processes physicians and surgeons must 'do no harm'. Three hundred years after Hippocrates, Dioscorides, a Greek physician formulated over 600 remedies from plant and animal organs, his medicines being prescribed over the next 1,600 years. During the second century CE, Galen of Pergamum added more remedies, known as 'galenicals', to Dioscorides' *Materia medica*.

It is impossible to discover what useful drugs the alchemists may have discovered in their pursuit of the *Elixir vitae*, the elusive substance that would ensure eternal life. The alchemists wrote their reports and results in cryptic codes and obscure symbols in order to confuse their competitors.

During the Middle Ages, Arab scholars achieved remarkable progress in science and the arts, in alchemy and medicine. In the eleventh Century CE a Mesopotamian physician, commonly known as Avicenna (his actual name was Abu Ali ibn Saud), translated the published works of Dioscorides, Galen, the School of Hippocrates and other Greek physicians into Arabic. Avicenna and his fellow Arab scholars also translated the writings of Aristotle and Plato, as well as other Greek philosophers and authors into Arabic. Subsequent translations into English and other European languages made possible our present access to Greek medicine and literature. Arabian *Materia Medica* of the period describes more than 1,000 drugs derived largely from medicinal plants.

From empiricism to science

From the mid-1800s analytical chemistry, microscopy and cytology progressed impressively. Anton von Leeuwenhoek, a Dutch draper, with no training in optics or any other scientific subject, learned to grind glass lenses and constructed a microscope with a 30x magnification by which he identified protozoa (which he named 'animalcules'), described blood corpuscles and the cellular structure of fish tissues. Chemotherapy, using chemical substances to inactivate disease-causing pathogens by bactericidal or bacteriostatic actions, was stimulated by identification of microbial pathogens and chemical means of controlling them. Wöhler's conversion of ammonium isothiocyanate into urea demonstrated that naturally occurring organic substances can be synthesised from non-biological sources. The discovery of hormones extracted from endocrine and ductless glands and later synthesised, added another dimension to therapeutic medicine, to the gradual development of modern pharmaceutical industries.

Despite some 20 per cent of modern pharmaceuticals being derived from natural and genetically modified organisms, a lively consumer and commercial interest in traditional sources of natural medicines persists and may be growing. Ethnobotanical expeditions in tropical forests of the Amazon have delivered more than 10,000 species for pharmacological examination. A vast unexplored resource of plants and microorganisms exists across the planet. Of more than 100,000 identified species, fewer than 200 microorganisms provide substances applicable in food and pharmaceutical industries. Higher orders of terrestrial plants represent more than 65 per cent of the Planet's biomass but fewer than 6 per cent of identified species are cultivated commercially. Of 80,000 vegetative species believed edible, barely 15 provide 90 per cent of human caloric intake.

The avaricious depredation of tropical forests by loggers and planters of cash crops, the despoliation of sites of rich natural biodiversity, deprive humanity of vast unexplored sources of medicinal plants and other potentially valuable organisms. Sadly no international agency exists with authority to prevent such predations.

Synthetic drugs and chemotherapy

Around the middle of the nineteenth century a chemist in London, William Henry Perkin, was trying to synthesise quinine, a drug used to treat malaria, traditionally extracted from the bark of *Cinchona* species indigenous to South American forests. Legend has it that, while heating a mixture of aniline and potassium dichromate, Perkin broke his thermometer, the liberated alcohol reacted with other substances to produce a rich purple pigment which he named 'mauveine', believed to be the first of the aniline dyes. Based on this result von Hoffman, Perkin's tutor, started the German dyestuff's industry. Encouraged by success with synthetic dyes, German companies Bayer, Hoechst and Merck began chemical synthesis of drugs, first by producing analogues and derivatives of therapeutically active substances extracted from medicinal plants [Hulse 2004].

In the seventeenth century an inquisitive English physician discovered that rural people in the midland counties drank infusions of willow bark (*Salix spp*) to alleviate arthritic and rheumatic pain. The active principle in the bark was later analysed and identified as salicylic acid. Chemists at the Bayer Company reacted acetic anhydride with salicylic acid to produce acetylsalicylic acid, the first proprietary synthetic drug 'Aspirin'. Other synthetics, such as codeine produced by methylation of morphine, soon followed. In 1936 the British Medical Research Council defined 'chemotherapy' as medical treatment by synthetic chemical compounds that react with specific infective organisms.

Antibiotics

While engaged in microscopic studies to discover what microbes were devastating silkworms, Pasteur suggested that microorganisms might be induced to attack other microbes. In 1928 Alexander Fleming in London observed that a mould spore from *Penicillium notatum* accidentally infected and inhibited growth in a bacterial culture. The therapeutic potential of Fleming's observation was ignored until re-examined in 1939 by Florey and Chain at Oxford University. The effective bacterial inhibitor was isolated, chemically characterised and named 'Penicillin', the first of a long series of antibiotics subsequently extracted from cultures of various *Actinomycetes* species. Long before Fleming's observation, anthropologists reported how Micronesians scraped mould from tree bark which they rubbed into wounds to prevent festering [Hulse 2004].

Hormones

More than a century ago Claude Bernard, a French physiologist, reported how some critical bodily functions are regulated and controlled by 'centres of internal secretion'. Later the active agents were identified as endocrine and ductless glands that secrete hormones, the name derived from ηορμον (hormon) a Greek word that means 'to urge on'. In the 1920s, adrenaline, extracted from suprarenal animal glands, was chemically analysed and industrially synthesised. In 1921, at the University of Toronto, insulin essential to diabetic patients was isolated from Langerhens Islets present in porcine pancreas. In the 1980s other Canadian scientists synthesised an insulin precursor from a genetically modified bacterium. Thyroxine, generated in thyroid glands, and various other hormones have since been chemically and/or microbially synthesised. Avian and bovine growth hormones, synthesised by genetically modified bacteria, promote growth in farm animals and cultured fish and stimulate lactation in bovines.

Food and pharmaceutical industries

Recognising their importance to human survival, healthy growth and activity, it is not surprising that food and drug industries constantly expand and diversify to satisfy changing demands of expanding, affluent and ageing populations. The total world value of industrially processed foods is estimated at more than $3 trillion, the sales value of commercial pharmaceuticals, excluding veterinary medicines, exceeds $500 billion, of which 49 per cent are sold in the USA, 24 per cent in the EU, 16 per cent in Japan. During the past 40 years food processing industries in India have grown to employ over 3 million people, to supply processed foods to more than 300 million regular consumers. The estimated value of foods processed in India in the year 2000 was more than 1,000 times the value in 1960.

Influence of supermarkets

Supermarkets command exceptional influence over agricultural producers and food processors. Supermarkets, whose shelves may carry 10,000 distinct products and which supply some 90 per cent of all foods purchased by consumers in North America and the EU, exert pressures on growers and processors to meet the supermarket's quality standards, standards often difficult for small producers and processors to satisfy. Because of high sales volume, supermarkets can offer prices lower than small retailers. Though among poorer nations most food is sold from public markets, local stores and street vendors, supermarkets, many subsidiaries of European or American giants, are now rapidly expanding in Asia and Latin America.

Post-harvest loss

To intellects of limited perspective, the primary purpose of food processing is to reduce what is variously described as 'post-harvest loss', 'degradation', 'deterioration', 'damage' or 'spoilage'. Post-harvest deterioration or spoilage, caused by biochemical change, microbial infections, insect infestations, attacks by other pests, chemical and other contaminations, can range from minor bruising or discoloration to foods heavily infected by pathogens or so spoiled as to be offensive and obnoxious. Post-harvest loss, of minor or major proportions, may involve biological, nutritional, economic or physical changes, result in lower weight, impaired appearance that is unacceptable, food unfit or unsafe to be eaten.

Assessments of post-harvest deterioration, degradation, damage, spoilage or loss, should state precisely what damage has been inflicted, the extent, cause, probable consequences and recommended means of remediation. Fully to understand the cause of post-harvest degradation it is essential to study systematically the entire sequence through which the product has been conveyed, to determine where, how and by what the damage was caused, a process that demands the competence of food systems analysts [Bourne 1977, FAO 1983, NAS 1978, Hulse 1995].

Industrial biotechnologies: changing patterns and future prospects

Research and development

Though similarities between food and drug industries, and in lesser degree with textiles, are noteworthy, many divergent differences are apparent. Small-scale apothecaries that compounded medicines, dispensed pills and ointments have been driven out by huge drugstore conglomerates that dispense pre-packaged prescription and non-prescription drugs together with a vast assortment of toiletries and cosmetics. Pharmaceutical industries that develop and manufacture drugs are composed of very few, large highly capitalised corporations, while food processors include giants such as Nestlé and Unilever as well as many thousands of relatively medium and small enterprises. Recent years have witnessed the creation of several thousand specialist 'biotechnology research companies', which, under contract or licence, develop new diagnostic, prophylactic and therapeutic substances and provide other technical services to pharmaceutical companies.

Pharmaceutical companies invest between 10 and 18 per cent of revenues in research, specialist biotech companies more than 30 per cent, and food processors an average of less than 0.5 per cent. A survey conducted in Canada a few years ago reported that of more than 3,000 registered food processors fewer than 20 maintained a liberally defined research activity. Only large food companies support a discrete research department, most others rely on raw material and equipment suppliers to provide technical advice. Some governments support national food research laboratories, which may offer technical advice to small industries. Many government research institutes cannot maintain a detailed awareness of specific opportunities, resources and constraints among hundreds of small and medium food companies. One encounters government food research institutes where few if any of the technical staff have experience in food processing industries. Consequently, while their scientific knowledge may be impressive, they possess little understanding of market forces, and opportunities. They therefore devise new products for which no consumer demand or profitable markets have been identified.

Biomedical industries

The biomedical industries are developing into two inter-related entities: a. large pharmaceutical corporations, many of which operate internationally; b. specialist bioresearch enterprises, sometimes described as 'second generation biotechnology companies' or more simply 'Biotechs' or 'Biotech/Biopharm'.

Biotechs started about 30 years ago by using genetic modifications to synthesise proteins and derived drugs. In 1970 two Americans, Cohen and Boyer, demonstrated how using restriction enzymes DNA could be cut and transferred between unrelated organisms. In 1976 Boyer with several of his Californian colleagues created Genentech, now described as the first modern biotechnology company. Optimistic speculation that genetic transformations would quickly generate valuable new diagnostics and therapeutics encouraged enthusiastic investment in the new biotechnology enterprises. When Genentech made its first public stock offering its share price rose from $35 to $89 in twenty minutes, the fastest rise in American stock market records. Subsequently the fortunes of Genentech and other new biotechs fluctuated as investor expectations waxed and waned. Some failed, many survived: around the world there are over 4000 biotech/biopharm research companies. Over 70 per cent of Canada's so-called biotech/biopharm research companies were spun off from universities.

During their early years biotech/biopharm companies were relatively small and confined to laboratory syntheses and micro-processes. Some in North America have diversified to engage in preclinical and early clinical trials of potential new drugs. Among Canadian companies that describe themselves as Research Biotechs, 75% are small with fewer than 50 employees, 10 per cent are relatively large with over 150 employees. More than 50 per cent are engaged in the human health sector (therapeutics and diagnostics).

A growing group specialises in bio-informatics: data bases and information that they sell to other companies. Some provide analytical services, others molecular libraries from which samples of particular organic composition can be supplied to clients. Since they began as small bioscience research companies, Biotechs have diversified into specialised diagnostics, biopharmaceuticals, food and agriculture, renewable energy, environmental protection and clean-up, protection of biodiversity, bioinformatics and biotechnology systems.

Many Biotechs spun off from universities have failed. Some lacked industrial management and marketing experience. Others began with too little working capital to survive the long R & D period before a profitable product or service is developed.

Drug development

The sequence of modern drug development begins by identifying targets for therapeutic intervention. These may include cellular or pathogen-specific molecules such as proteins critical in a targeted disease. Compounds may also be discovered from several thousand accessions in molecular screening libraries. Identified compounds are refined and subjected to a battery of biochemical analyses, simulation models and tests with laboratory animals. Compounds that pass preclinical screening proceed to a series of controlled clinical trials with human subjects. The high costs of clinical trials are forecast to increase at over 12 per cent a year. Drugs that pass clinical trials are presented for evaluation and approval by regulatory agencies. Highest risk and rate of failure occur during early preclinical evaluations. In North America only one out of several 100 compounds initially examined survives clinical trials;

only 20 per cent of those rated clinically acceptable find profitable markets.

In December 2004, patents on some 15 highly profitable drugs expired. Consequently there is pressure on biotech/biopharm and pharmaceutical companies to discover and bring to market effective new drugs of wide potential application. Pressures on drug companies are further exacerbated by rising demands from ageing North Americans and Europeans for pharmaceuticals, prosthetic devices and other health-care services. Drug companies are on the horns of a dilemma: people suffering from serious maladies, such as HIV/AIDS, demand rapid approval of new therapeutics. Fearing costly litigation when new drugs cause unforeseen adverse effects, drug companies prefer to proceed cautiously and patiently until all clinical trials are satisfactorily complete.

Several Biotechs that began specialising in genomics, combinatorial chemistry, specific drug discovery or monoclonal antibodies to be licensed to larger companies, now carry new discoveries to pre-clinical, some to early clinical trials. Successful companies have defined their objectives according to their resources and concentrated on specific drugs for widespread diseases, or on technological systems ('platforms') amenable to several applications. Some specialise in diseases related to specific genes and means to remedy genetic dysfunctions.

To gain an adequate critical mass of resources, biotech research companies are acquiring or merging with competitors. Others enter into alliances with pharmaceutical corporations, some by mergers, others through formal licensing arrangements. Over one-third of American biotech research companies depend on licensing products and processes to larger biotech or pharmaceutical companies.

The human genome

Elucidation of the human genome sequence stimulated investment in pursuit of diagnostics and therapies for genetically induced maladies. Early expectations were modified when it was discovered that the human genome contains about 30,000 genes – not, as originally believed, close to 150,000 genes. Each active gene may encode for several different protein molecules, so there may be 5 times as many drug targets as there are genes. Recent speculations variously suggest that of the present estimated total of *ca* 500 targets, genomic research will lead to eventual identification of between 3000 and 10,000 targets, corresponding to proteins encoded by between 10 and 25 per cent of identified human genes. Following on Dolly the sheep, there is concern that knowledge derived from the human genome sequence could be adapted to clone humans, a practice some would equate with man trying to play God.

The ability to sequence genomes has stimulated an intriguing diversity of biotechnological sub-sectors. *Genomics* includes the study of the structure of genomes, genomic functions related to specific diseases. *Metabolomics* is devoted to the investigation of how metabolites influence physiological dysfunctions. *Pharmacogenomics* studies genomic and genetic patterns and malfunctions related to specific diseases, to molecular structures as they influence pharmacological activities. *Proteomics*, a rapidly expanding sub-sector, is concerned with specific proteins as they relate to particular infections, to pathogenic modes of action, proteins as carriers for vaccines, genes that encode for novel enzymes. Scientific institutions in some 18 nations are collaborating members of the Human Proteomics Organisation (HUPO), coordinated from a headquarters located at McGill University in Montreal.

In a recent lecture, Dr Elias Zerhouni, director of the US National Institute of Health, described how biomedical research and applications are changing from design and synthesis of pharmaceuticals to alleviate diseases to interventions at the molecular and cellular levels that aim better to understand the cause and etiology of various diseases, some communicable and others brought about by genetic dysfunctions.

Nano-sciences

Nano-sciences study the molecular constituents of biological substances and materials. 1.0 nanometre (nm) = 10^{-9} metre; the diameter of a human hair ranges between 100,000 and 200,000 nm. Novel analytical instruments, including the Scanning Tunnelling Microscope and the Transmission Electron Microscope facilitate determination of the structure of molecules. The properties of biochemical macromolecules are significantly influenced by the nature and structure of their constituent molecules. It is postulated that, by designing and synthesising novel molecules of desirable structure and properties, the functional behaviour of many proteins, pharmaceuticals and eventually food materials can be made more efficient and more effective [Roy Soc, 2004]

Costs and revenues

Between 2001 and 2002 global revenues of Biotechs rose from $35.9 billion to $41.3 billion, of which 73 per cent was earned by American companies. In 2001, total revenues of the six largest bio-research companies amounted to about $8 billion, their research and development investments were between 20 and 30 per cent of revenues. These specialist Biotechs devise and develop new drugs and processes which some carry to preclinical stages of safety and effectiveness, and pilot-plant production scale. Larger pharmaceutical corporations expand the smaller biotech processes and subject the products to more intensive and extensive *in vitro* and *in vivo* clinical trials to determine potency, reliability and safety before submitting the new drug, with all available relevant scientific evidence, to regulatory authorities for official approval.

Pharmaceutical companies are often criticised for their high prices and revenues. Critics ignore the exorbitant cost of developing a successful new drug, now in excess of $800m increasing at over 10% year, the time required from early laboratory synthesis being up to 15 years. Fewer than 1 per cent of substances initially synthesised are eventually approved, most being rejected at various stages along the way. The patent on a new drug lasts for only 20 years from the time the patent is approved. As soon as a patent expires, companies that are designated generic companies start to produce and sell the drug. Generic companies can sell at a lower price, since they incur no lengthy research and development costs. Some regulatory authorities have established departments specifically responsible for the monitoring of drugs processed by generic companies, to ensure that the generic version is equal in composition, potency and safety to the original.

To economise, several pharmaceutical companies have merged with competitors, others are 'downsizing'; Pfizer will soon close seven factories and terminate several hundred employees.

Bio-informatics

An early seventeenth century scholar complained: "The disease of this age is the multiplicity of books; we cannot digest the abundance of written matter." How to record, classify and provide access to the overwhelming volume of bio-scientific knowledge now generated deserves serious, systematic attention. More than 350 bio-informatic databases operate around the world. Urgently needed are bio-scientists and bio-philologists competent to compile and maintain reliably accurate databases and bio-information systems using precise terminologies comprehensible to all concerned. Database curators need competence to recognise accuracy and relevance, to delete outdated accessions, and interpret needs of potential users. Several companies maintain molecular libraries that offer computerised molecular structures and provide samples of the substances illustrated in their computers to research biotech and pharmaceutical companies. An urgent need of bio-informatics is for an internationally accepted thesaurus and dictionary of precise definitions. The penchant to coin new names without stating precise definitions or how they may conflict with existing terminologies persists as a serious impediment to rational databases and cogent bio-scientific communication.

Industrial biotechnologies: trends and prospects

Semantic confusion

Any attempt to suggest how the future of industrial biotechnologies may unfold is hampered by confused concepts and imprecise interpretations of what is connoted by 'biotechnologies', with no clear distinction between 'ancient' and 'modern' technologies. Disturbingly typical is a publication by the Directorate-General for Press and Communications of the European Commission [Eurobarometer 2003]. The publication is the fifth in a series of reports of surveys to assess the attitudes of Europeans to 'biotechnology'. The report speaks of 'modern biotechnology' without any definition of its intended meaning. It differentiates between 'GM Crops' and 'GM Foods' seemingly unaware that many GM foods are derived from GM crops, some with ingredients produced by GM organisms. GM foods are described as 'using modern biotechnology in the production of foods, for example to make them higher in protein, keep longer or change the taste'. GM crops are described as 'taking genes from plant species and transferring them into crop plants to increase their resistance to insect pests'. Foods may be preserved, processed or transformed by modern biotechnologies that do not involve genetic modifications. Is a food partly processed by an enzyme, or with a flavour or micronutrient derived from a GM organism a GM food? When does a modern biotechnology cease to be modern?

Biotechnologies employ a diversity of cell cultures. Various metabolites applicable in food processes are synthesised and extracted from cultured cells of microorganisms, higher plants and other organisms that have not been transgenically modified. The EC report states that one-half of the people surveyed were asked for their opinions on 'biotechnology', while the other half were questioned about 'genetic engineering'. In the final analysis the results from the two groups were combined, clearly indicating that in the minds of the surveyors 'biotechnology' and 'genetic engineering' are one and the same. By rational scientific interpretation, 'genetic modifications' are simply components of a very broad and constantly expanding spectrum of 'biotechnologies'.

Over the past 20 years, biotechnologies have evolved from intellectually intriguing biosciences, into industries that process and produce biological products from bio-catalytic reactions, from natural and GM microorganisms, plant cells, mammalian and insect cells. Some techniques modify genetic composition and expression, others accelerate and adjust metabolic processes.

Cell cultures complement chemical modifications in syntheses of diagnostics and therapeutics. Cultured microbial and viral cells, some genetically modified, produce a diversity of hormones, nutrients and drugs. Insect cells infected with adeno-and baculo- viruses produce proteins used in gene therapy and viral vaccines. Mammalian cells synthesise therapeutics of superior purity: monoclonal antibodies, viral vaccines and Factor VIII to promote blood clotting in haemophiliacs.

Cell cultures demand qualified bioengineers and systems analysts to rationalise interactions among many complex variables in the selection of media, variable conditions of culture, all of increasing complexity as culture processes progress from laboratory, through pilot plants to factory-scale.

Pathogenic bacterial colonies transmit biochemical signals to determine when they have sufficient umbers to overcome a target host's immune defences. They then turn on virulent genes to become potently infective. Scientists in Wisconsin have developed derivatives of L-homoserine lactones to block bacterial signals rendering the pathogens susceptible to antibiotics.

Molecular modelling and virtual reality

Pharmaceutical industries began by extracting, screening and chemically modifying pharmacologically active substances of natural origin. These traditional processes are giving way to identification of how specific diseases are caused, the mechanisms by which particular drugs act to diagnose, prevent or cure them. The latest generation of prophylactics, therapeutics and diagnostics are designed by molecular modelling in programmed computers. Late in the last century, using classical organic chemistry, a scientist could synthesise about 50 new active compounds in a year. Today, computer assisted biosyntheses can generate several thousand new compounds. Programmed computers devise molecules to be systematically compared with molecular structures stored in the computer memory or downloaded from molecular libraries. In less than a year a million compounds can be screened against specific target proteins.

Other biotechnological innovations include 'Genomics' – the study of genomes and DNA nucleotide sequences; 'Proteomics' that study specific proteins produced by genomes; 'Metabolomics' that study the influence of gene expression on metabolites synthesised in cell cultures.

'**Virtual Reality**' provides novel means to study molecular structures, DNA, whole organs, plant and animal cells. At the University of Calgary, Java 3-dimensional technology projects images onto the walls and floor of a cubic room 2.5 × 2.5 × 2.5m. The observer virtually stands in the middle of the molecule or human organ being studied. Images examined through specially designed spectacles can be expanded or moved around by a manual control. Trained observers can study 3-dimensional structures of DNA to discover aberrations or unusual features; the structure of whole organs; the structural relation between a synthesised diagnostic or therapeutic and a diseased cell.

Downstream processes

In biotechnological nomenclature, 'downstream' relates to all processes that follow synthesis in a bioreactor: fractionation, isolation, purification and sterilisation of synthesised products. Most difficult and expensive are scale-up processes: expansion from a laboratory flask, through pilot plant to processes of industrial scale. In North America, downstream processes absorb some 80 per cent of factory production costs. Highly skilled bioengineers are in urgent demand to improve the efficiencies and economies of downstream processes. Recent studies of skilled human resource needs indicate that present and predicted demand for biotechnologists and bioengineers significantly exceeds expected supply.

Separation and isolation

Different primary products are isolated from different fractions: insulin from harvested cells; several vaccines from supernatants; intracellular metabolites are released by rupture of cell walls. Substances sensitive to excessive heat, organic solvents and other chemicals, need benign systems of fractionation, isolation and sterilisation. In bio-chemicals sensitive to heat, microorganisms can be inactivated by high-temperature-short-time [HTST] sterilisation, or removed by microfiltration. Intact cells can be harvested by continuous centrifuging, but carefully controlled conditions are needed for sensitive mammalian cells. Cell disruption to isolate intracellular metabolites can be mechanical, chemical or enzymatic. Mechanical methods include high-pressure homogenisation and maceration using small glass or plastic beads. Enzymatic lysis of cell walls are conservative of energy and less prone than mechanical or chemical methods to degrade sensitive bio-synthates. Processes catalysed by enzymes, many derived from GM organisms, being less demanding of energy and more benign than many chemical catalysts, are finding increasing bio-industrial applications.

Optimum conditions for isolation and purification of metabolites vary with the substances involved. Liquid extraction is useful for antibiotics; fractional distillation for heat-tolerant organic chemicals; freeze-concentration for heat-sensitive substances in aqueous dispersion, aqueous-phase liquid fractionation for heat-sensitive enzymes. Membrane separations include reverse osmosis, ultrafiltration, microfiltration, electrodialysis and nano-filtration. Ancient Egyptians applied microfiltration to purify water from the Nile that slowly percolated from settling ponds through fine sand filters.

At pressures between 10,000 and 40,000 kPa carbon dioxide behaves as an organic solvent and is used to extract bio-chemicals intolerant of organic solvents. Supercritical gas and liquid extraction [SGE] is used industrially to isolate essential oils and oleoresins from aromatic plants, and caffeine from coffee.

Processes of preservation

Most foods and pharmaceuticals are susceptible to deteriorative change: through microbial infection, pest infestation, biochemical reaction, chemical contamination and atmospheric pollution. Sensitive foods and pharmaceuticals can be preserved by processes that

a. inhibit, destroy, remove or prevent entry of harmful microorganisms
b. restrict adverse biochemical and biophysical change.

Biochemical spoilage induced by microbial infection or catalysed by enzymes can be restricted by

a. thermal destruction or inactivation of an offensive microbe or enzyme
b. reducing temperature of the product and ambient atmosphere to below freezing
c. lowering the active water content (i.e., product vapour pressure) by partial dehydration or elevation of soluble solids
d. control of gaseous ambient atmosphere within the package or container.

Reducing the active water

Active water $[A_w]$ is defined as the quotient of the vapour pressure of a particular biological material divided by the vapour pressure of pure water. Reducing Active Water restricts microbial growth and biochemical degradation since most enzyme catalysed degradations require an aqueous environment. Active water may be reduced a. addition of soluble solids, the A_w of saturated salt or sugar solutions are significantly lower than the A_w of pure water; b. removal of constituent water by processes of dehydration.

Partial dehydration of harvested cereals, meat, milk, fruits and vegetables by radiant heat from open fires or by exposure to direct sunlight has been practised for many hundred of years. More benign dehydration systems, that inflict less thermal damage, rely upon shorter times of exposure to heat, evaporative cooling, lower temperatures and reduced ambient pressures to accelerate rates of aqueous evaporation. Dehydration under vacuum depresses the boiling point of water and reduces ambient resistance to evaporation.

Lyophilisation [freeze drying] is protective of biological materials highly sensitive to thermal damage. Most biological cells are ruptured by large ice crystals that form during slow freezing. Lyophilisation begins with rapid cryogenic freezing by immersion in a medium below-150°C, ice crystals, rapidly formed are small and do not rupture organic cells. The frozen material is enclosed in a vessel sufficiently robust to withstand an internal pressure below 5.2 mbar under which conditions the material remains frozen. The constituent ice sublimes directly into water vapour which may be captured on a condenser. Rates of sublimation can be accelerated by black surfaced heating plates or by micro-wave heating.

Controlled atmosphere storage

Perishable foods, sensitive to oxygen or other ambient gases, may be stored and/or packaged under partial vacuum or an inert gas. Deterioration by oxidation can be reduced by storage/packaging under partial vacuum, nitrogen or carbon dioxide. Many fruits synthesise and exude gaseous ethylene which accelerates ripening and reduces shelf-life. Bananas stored under an atmosphere containing 5 ppm of ethylene will ripen in 2–3 days. Storage under an atmosphere high in carbon dioxide and relatively free of ethylene will delay ripening for 16–18 days.

Thermal and non-thermal preservation

Thermal processing can impair critical functional and physiological properties in addition to inactivating undesirable micro/organisms and enzymes; the higher the temperature, the longer the exposure, the greater the degree of damage. Food processors are faced with a challenging dilemma. During the past 30 years, reported food-borne infections have increased by several orders of magnitude, pathogens such as *Campylobacter, staphylococci,* novel strains of *E. coli* and *Listeria* recently added to the long-time nuisance pathogen *salmonella*. Consumers now demand foods perceived to be healthier, fresher, and less perceptibly altered from their natural state.

Industrial technologies are designed to inactivate or slow down microbial and biochemical processes of spoilage and degradation. To restrict thermal damage, various relatively benign thermal and non-thermal processes are under investigation. Processes with short exposure times include spray-drying, tubular and scraped-surface heat exchangers, steam injection before aseptic packing. Aseptic systems rely on physical exclusion of microorganisms and contaminants from sealed sterile systems by which processed products can be conveyed or stored. Thermal damage can be restricted by uniform generation of heat for a relatively short time.

Ohmic heating is effective for biological materials sufficiently fluid to be pumped through a column across which an electric current passes between two electrodes. Heat is generated rapidly and uniformly, the sterilised product is rapidly cooled, aseptically conveyed and packaged in sterile containers. Commercial models range from 10 kW pilot units that process 100 kg per hour up to 300 kw machines with capacities of 3 tonnes per hour.

Microwave (MW) and *radio frequency* (RF) heating depend on electromagnetic energy from a magnetron that produces an electric field alternating at microwave and radio frequencies. Heat is generated by rapid reversal of molecular polarisation. MW and RF provide uniform short-time heating at high internal temperatures. MW and RF are used industrially in dehydration, microbial inactivation and rapid cooking, with optimum processing parameters determined by computer simulation models.

Non-thermal processes

These include *high isostatic pressures* between 50 kPa and 1 MPa applied to biological materials high in water content. Products are sealed in evacuated flexible packages immersed in fluids; high pressures are uniformly transmitted through the fluid medium to the packaged materials. Ultra-high pressures cause temperatures to rise in most compressible substances, the degree of increase depending on the specific heat and other physical properties. Heat generated in compressed gases provides the operating principle of diesel engines. Materials high in water suffer measurable but lower temperature increases than compressed gases. Under adiabatic conditions near to normal temperatures, temperatures in high moisture foods rise in linear relation to the pressure applied, temperature rise among different foods ranging from 3° to 6° for every 100 MPa pressure increase.

Three forms of *pulsed energy* for microbial inactivation are under study: a. Pulsed electric fields (PEF); b. Pulsed light (PL); c. Pulsed magnetic fields (PMF). Electric potentials induced by PEF cause lethal polarisation of microbial cell membranes, empirically determined critical potentials varying among microbial species, cellular morphologies and ambient conditions. Alternating pulses seem more effective than constant polarities at field strengths between 15 and 30 kv/cm. Both PEF and PL, the latter generating light flashes at 20,000 times sunlight intensity at the earth's surface, are effective against vegetative microbial cells, but not against spores or active enzymes. Oscillating magnetic fields are said to inactivate vegetative microorganisms though the literature on industrially viable applications is sparse.

Ultrasonics at frequencies above 20 kHz, causing alternating pressures and cavitation, can inactivate vegetative microorganisms in liquid suspension. Ultrasonics may also be used as an adjunct to crystallisation, filtration, hydrogenation and ageing of alcoholic beverages.

Ionising radiations can inactivate microorganisms and destroy insect infestations. Radiation sources include gamma rays from such radioisotopes as ^{60}Co and ^{137}Ces; X-rays or electrons generated by machines. Absorbed radiation is measured in Grays or Kilograys (kGy), one Gray equivalent to one Joule/kg. In the United States irradiation is permitted to preserve various biological materials. Permitted maximum doses are indicated parenthetically for the following: fresh poultry (3 kGy), dehydrated enzymes (10 kGy), spices (30 kGy), pharmaceuticals (above 30 kGy dependent on biochemical nature and conditions of use). Higher levels are allowed in food ingredients consumed in small quantities (e.g., spices) and in prescribed pharmaceuticals.

Process and product quality control

Simply defined, quality control (QC) objectives are to ensure properties of raw materials and final products comply with defined specifications and specified properties are consistently maintained in all products processed over time.

QC specifications are defined by: a. international protocols, b. government regulatory agencies, c. customers, secondary processors and retailers, d. food and drug processing companies.

For many years product QC was determined by extracting representative samples of processed products to be assessed and analysed in quality control laboratories. *Ex post facto* laboratory analyses are gradually being superseded by on-line sensors and monitors that continually assess and detect defects in critical properties. Feed-back signals correct faulty processing parameters, all on-line determinations being recorded in computers. Sensors variously depend on biological, chemical, biochemical, biophysical, microbial, electronic and photoelectric signals. They may record apparent viscosities, temperatures, pressures and relative humidity. Chemical sensors respond to pH, specific ions, organic radicals and impurities. Biosensors may employ immobilised enzymes, bacteria, antigen-antibody reactions and DNA probes. Spectroscopic methods, too many and varied to be comprehensively described, respond to physical and rheological properties and to many chemical constituents.

Magnetic resonance imaging (MRI), generally too expensive for food processing but used by pharmaceutical industries, depends on the different magnetic properties of atomic nuclei placed in a magnetic field that induces different energy levels between protons aligned with and against the field. MRI can detect many defects and being non-destructive, installed on-line it can assess 100 per cent of processed, packaged products, and is particularly useful in monitoring packaged pharmaceuticals.

On-line sensors must be accurate and reliably responsive, tolerant to processing conditions, easy to install and maintain. Sensors and their responses must be systematically checked and correlated with direct analytical determinations. It is reported that up to 1000 different nano-sensors can be carried on a single on-line sensor-detector.

Final comment

The foregoing illustrates how industrial biotechnologies have progressed from perceptive artisanal empiricism to scientifically devised and controlled systems. Scientific development and control are more advanced among pharmaceutical than among food and textiles industries, where scientifically devised systems continue side by side with traditional, empirical technologies. Supermarkets and other dominant retailers compel food and textile industries to deliver goods at ever lower prices. Low prices are attractive to consumers but present trends forebode that only large food industries can afford the research needed to develop and deliver products of consistent quality at prices demanded by supermarkets. Eventually this could result in significant reduction in the diversity of foods available and affordable to most consumers and the elimination of many small-scale farmers and retailers. Large department stores and supermarkets that seek to undercut the prices of competitors for textile goods are stimulating production by low paid workers, some in poor countries, others in oppressive sweat shops staffed by immigrants.

Given the high cost of modern biotechnology research and development, it is doubtful if the innovation, design, production and distribution of all necessary foods and essential drugs can be controlled solely by market forces. Government interventions are inevitable if all humanity is to be assured of food and health security. Mortality and morbidity from tropical diseases will be exacerbated by global warming; pandemics will spread across the planet by intercontinental migration of humans, birds and by goods transported in international trade. The need for progressive research to expand and refine biotechnologies is a *sine qua non*, for as René Descartes observed in his 17th century *Discourse on method*: "All at present known is nothing in comparison with what remains to be discovered."

10

Environment and Resources

The planet's creation and evolution

The noun 'Environment' is derived from an ancient French word meaning 'to surround' or 'to encircle'. 'Environment' is defined in the *Oxford English Dictionary* as 'the surrounding conditions or influences under which organisms, persons or things exist and develop'. From this broad definition 'environment' is applied to various physical, political, economic, academic, social systems and conditions. This text is confined to the Planet's physical and biological environment, its atmosphere, climates, ecosystems and natural resources and their relation to biotechnologies.

Anthropological, cosmological and geological evidence indicates that from the time of its creation the Earth, its environment, ecologies, natural resources and the creatures and organisms that have been its inhabitants have continually changed, sometimes by gradual evolution, sometimes from extreme and disruptive occurrences: collisions with large asteroids and meteorites, violent earthquakes, volcanic eruptions and extreme climate change. Justifiable is a growing concern with undesirable changes accelerated by human activities: gaseous and particulate emissions from combustion of fossil fuels and timber; agricultural, industrial and domestic pollution of atmospheres, inland and coastal waters; deforestation; degradation of land and ecosystems; deliberate or consequential depletion of biodiversity. A comprehensive review of how the universe and planet earth were created and have evolved, the origins and evolutions of living organisms is recorded in a special issue of *Scientific American* [*Scientific American 1994*].

An increasing number of reputable observers contend that the styles of more affluent middle classes, now expanding as countries such as the People's Republic of China and India rise in prosperity, with dilating demand for motor vehicles and other amenities, cannot be sustained with the resources available. Resources are being used at a faster pace than they can be restored. Human-induced modifications to the environment, to the global biochemical and hydrological cycles are rapidly accelerating. "Most transformations over the past 10,000 years have occurred during our lifetimes" [Kates 1994].

Governance of environments and resources

The United Nations Environment Programme, established following the 1972 Stockholm conference, conducts research and surveys related to environmental issues. Other UN agencies and international organisations study and report on the Planet's diverse resources, provide estimates of consumption and depletion, and recommend protocols, procedures and practices to encourage conservative resource utilisation. Sadly no single international body possesses the mandated authority and responsibility to protect the Planet's environment, nor to control access to and consumption of its natural resources. Given the isolationist political ideology of the most powerful nation, creation of an international environmental authority empowered to enforce conservation seems improbable in the near future.

National, state and local governments can enact and enforce laws related to resources within their territories and jurisdictions, for control of atmospheric and other forms of pollution, for disposal of sewage, hazardous and other wastes. Governmental legislative protocols and regulations vary among nations, many persuasively influenced by industries and organisations with vested interests. Conformity to such international conventions as relate to reducing atmospheric pollution, protecting biodiversity and overexploitation of marine resources is voluntary and no nation can be forced to ratify or comply with what is proposed. Among abstainers to convention conformity are the most egregious polluters and consumers, both governmental and industrial, of non-renewable resources.

Significant inconsistencies among nations are apparent anent ownership of and access to inland waters, natural forests and public land. Nor is there uniformity in how values of natural resource assets are assessed and accounted for. In some jurisdictions, trees have no declared value until they are chopped down; inland waters are free to everyone until they are pumped and piped to homes, industries and farmers. The failure to charge a rational price for water delivered to farmers leads to excessive consumption. Timber companies that clear-cut natural forests to extract only a few prime species are rarely charged for the unused timber left to rot. In several Latin American countries it was discovered that the trees extracted and used by lumber companies represented less than 15 per cent of the timber hacked down. There is no international authority empowered to force industries or individuals, whose activities inflict damage, despoliation or pollution on other nations, to reimburse their neighbours for the damage they suffer.

It would seem, therefore, that environmental protection and resource conservation are a collective, uncoordinated responsibility among individuals, communities, industries, state and national governments. Ideally all parties would cooperate with counterparts in other nations, with sensitive concern for the Planet's resources and the welfare of all its inhabitants. Sadly those nations and communities that consume most, who cause greatest environmental disruption and resource depletion, seem least disposed to exercise constraint and prudent conservation; they assume an inalienable right to increase their excessive consumptions.

Brundtland's plea to act with consideration for the needs of future generations falls more on deaf than receptive ears.

Are resources sustainable?

During the twentieth century world population grew from 1.6 billion to 6.1 billion. How rapidly and to what size it will expand during the twenty-first century defies prediction. In absence of some cataclysmic event or widespread disease epidemics, population growth, with persistent rising consumption of natural and industrially derived resources, will irreparably stress the Planet's resources and ecologies. Contradictions over potential consequences of ever-rising populations, resource depletion and environmental change are polarised between two extreme positions: (a) increasing population and resource demands pose a serious threat to environmental stability and finite resources; (b) human creativity and scientific ingenuity will devise technological means and resources by which to increase the Planet's resources and human carrying capacity. One school contends that scientific study should focus less on population growth and concentrate on assessing present and predictable increases in *per capita* consumption of renewable and non-renewable resources [Arizpe *et al* 1992]. The British New Economics Foundation, based on sound scientific evidence, states: "The Planet's ecological stocks are being depleted faster than nature can regenerate them." It seems evident that the Planet's essential resources cannot sustain predictable future demand and that either the more affluent must reduce their patterns of consumption or the poorest will suffer intolerable deprivation. In light of the serious and growing incidence of obesity and the many resultant diseases, it would seem desirable that all physically able human beings would be wise to use bicycles or walk more frequently than to travel by automobile.

Confusion of carrying capacity

Various soothsayers offer disparate predictions of the Planet's human carrying capacity, that range from 7.5 billion [Gilliand 1983] to

40 billion [Revelle 1976] to 50 billion [Brown 1954]. The concept of carrying capacity derives from animal husbandry: how many grazing bovines or ovines can be sustained on any given area of natural or cultivated pasture. Such estimates of global carrying capacity are of little practical utility. People are not grazing animals; they live, move and have their being in nations and communities, among and within which access to resources vary significantly. A more equitable sharing of worldly wealth, power and resources among all peoples, while highly desirable, seems unattainable given present egocentric dispositions. Global carrying capacity would differ greatly if all communities adopted America's level of resource consumption or everyone on earth adopted more conservational standards. Observers described by some as gloomy pessimists by others as pragmatic realists, profess beliefs that non-renewable resources, ecosystems, biodiversity, goods and services essential to healthy lives are being depleted, degraded, despoiled and over-taxed beyond economic or scientific restoration.

Ecosystems are the basic functional units of ecologies. They comprise a mix of abiotic materials and organisms that produce, consume and/or decompose other organisms. When undisrupted by external activities and influences, components of each ecosystem evolve and develop to co-exist in a relatively balanced symbiosis. If one or two components in an ecosystem are destroyed or depleted the balance is upset, other interdependent organisms disappear, exotics may arrive, biodiversity is deliberately and consequentially altered.

At the end of the twentieth century affluent OECD nations, that represented 16 per cent of the world's population, consumed close to 75 per cent of gross global product and almost two-thirds of the Planet's energy sources. Among Asian nations almost all available arable land is under cultivation, quantities of cereal grains harvested *per capita* have virtually levelled off, urban populations grow at 3.5 per cent a year, rural people migrate into cities, motor vehicles increase proportionately faster than human populations. 'Carrying capacity' is therefore more complex than a simple arithmetic relation between population numbers and the size of potentially arable land.

The people that on Earth do dwell

It is difficult to determine accurately the size of populations resident in any region or nation. Organisation of a national population census is costly and entails employment of many enumerators and statisticians. Complexities are compounded by shifting populations, migrations provoked by civil conflict, people seeking a better quality of life, by irregular dispersions of families and communities, some to places difficult of access. All quoted global demographic statistics must therefore be regarded as best estimates, biggest errors being among nations with least resources.

Blaxter presents an estimated pattern of world population development since settled farming began around the end of the Pleistocene period [Blaxter 1986]. Between 10,000 BCE and 01 CE the world's population is believed to have grown from 10 million to some 250 million; between 01 and 1800 CE from about 250 to 550 million. Population growth was never consistent or uniform; numbers were intermittently reduced by plagues and pandemics in the fifth and fourteenth centuries, by conflicts and famines in most centuries. In 1950, world population was assessed as 2.6 times the size in 1800; from 1950 to 2000 estimated growth was from 2.5 billion to 6.1 billion. How rapidly and to what eventual dimension world populations will expand during the twenty-first century and whether global population will eventually stabilise is highly speculative. Over the near future, urban communities will expand and become more dense and congested, patterns of food and resource consumption will change, stress imposed on all natural and industrially processed resources will be more destructive and depletive.

Existing inequitable conditions, with the rich and powerful acquiring greater wealth and power, while the poor and powerless suffer more abject misery, will exacerbate violence. Communities forcibly driven from their homes, dispossessed of their properties, deprived of their livelihood and condemned to near slavery, are inevitably motivated to attack those they regard as the responsible enemies. Whether those who attack their predators should be castigated as terrorists or hailed as patriots and martyrs, fighting and dying to recover their rights, is of conflicting opinion. As illustrated by the excessive damage and deaths inflicted upon Lebanon during the summer of 2006, opinions are entirely influenced by governments and peoples who support one or the other side.

Demographic patterns during the past millennia are quoted in Table 14.

Table 14
Population growth and life expectancy

Year BCE	World popn [M]	Year CE	World popn [M]
10,000	10	01	250
5,000	40	500	200
1,000	80	1000	250
500	150	1500	450

	Popn [Bn]			Growth	Annual % Increase	
	1950	1990	1995	2003	1950–2003	1980–85	1990–2003
World	2.52	5.28	5.92	6.24	×2.5	1.8	1.6
Africa	0.22	0.63	0.73	0.86	×2.7	2.9	2.7
Asia	1.40	3.20	3.46	3.73	×2.7	1.9	1.5
Latin America	0.17	0.44	0.48	0.55	×3.2	2.1	1.6
Europe	0.54	0.72	0.73	0.75	×1.4	0.4	0.1
Developing	1.79	4.27	4.67	5.14	×2.7		
Developing % of World	71	81	82	82			

Table 14a

	Life expectancy (Years)		<15 Yrs		Age groups % >65 Yrs	
	1975	2000	1975	2000	1975	2000
World	58	66	37	29	6	7
Africa	46	49	45	44	3	3
Asia	56	68	40	29	4	6
Latin America	61	69	45	44	4	5
Europe	71	74	24	17	11	15

Europe, in common with other affluent regions, is home to increasing numbers of older men and women, a rising proportion of non-producers with greater demands for health care and social services. In contrast, nations of Africa, Asia and Latin America include higher proportions of active young men and women who need employment and have larger appetites. If their native countries are unable to feed and sustain them, there will be pressures on more affluent nations to permit larger immigrations from poorer nations. The populations of Africa's poorest nations are predicted to double during the next 50 years.

Urban development

Christians and Jews, who accept the biblical accounts of creation as literal truth, believe the first humans appeared in the Garden of Eden. In contrast, Babylonians believed their god Marduk first created a city, which they called Eridu, and the first human beings to inhabit and rule it. Remains of Eridu were discovered by archaeologists in the mid-nineteenth century at a site known as Abu Sharein. The Judean-Christian mythology may have inspired the poet William Cowper to write: "God made the country, but man made the town." Since no one has discovered the Garden of Eden, while Eridu was inhabited for several hundred years, Babylonians may have reason to claim theirs as a more credible version of human creation. Eridu was located on the south bank of the Euphrates at the edge of the fertile alluvial plain, close to the southern marshes.

The most enduring legacy of the Babylonians was their invention of printing in Uruk and the planning of cities, of which there were several dozen during the last 3,000 years BCE. Each city controlled its system of governance, surrounding arable and pastoral territories, water management and irrigation. By efficient crop cultivation on the fertile alluvial soil, with supplementary irrigation from elaborate systems of canals and dikes, fed from the main rivers, Mesopotamian farmers produced food in excess of their subsistence needs. Surpluses provided food for the growing urban populations, enabling them to engage in non-agricultural profitable commercial and administrative activities.

The invention of cuneiform script in the city of Uruk about 3500 BCE, which eventually evolved into a hybrid of pictographic and phonetic signs, provided detailed records of how the Mesopotamian cradle of civilisation developed. Cuneiform is a method of writing in which a stylus cut from reed is pressed into soft clay to form wedge-shaped imprints. An assortment of patterns and designs depict different words and numerals [Latin '*cuneius*' = wedge]. [Leick 2002].

The ten most important Mesopotamian cities began with Eridu about 5000 BCE, and continued until Babylon, which lasted from around 2000 BCE until its decline during the Sassanian occupation between 225 and 600 CE. (The name 'Babylon' derives from the Sumerian 'Bab-Ilanl' meaning 'Gate of the Gods'.) When Alexander the Great defeated Darius in 331 BCE he intended Babylon to be the capital of his Persian empire. Had Alexander not died prematurely in 323 BCE, Babylon not Baghdad might now be the capital of Iraq. Baghdad, the city created by Al-Mansour, ruling Caliph from 754 to 775 CE, was the capital of the vast Abbasid empire. The city is located between the Euphrates and Tigris where the two rivers flow closest to one another. It was officially named 'Madinat as-Salam', 'City of Peace' but is known as Baghdad, name of the small village it displaced and superseded (Lewis 1993).

Various influences have led to the growth of cities and urban development around the world. Industrialisation in Britain caused textile spinning and weaving in rural homes to be replaced by textile factories powered by cheap coal. Countless unemployed rural people migrated into urban slums. Rapid mechanisation of American agriculture forced similar rural to urban migrations in the mid-nineteenth century. In many lands, migrations of young people into cities are stimulated by the attraction of amenities for entertainment and diversion unavailable in rural areas.

Unplanned urban growth

Unfortunately, most contemporary cities, unlike those of ancient Mesopotamia, have grown rapidly with little planning or foresight.

Older cities in North America and other territories now suffer from disintegration, breakdown, inadequate restoration and maintenance of infrastructures and services essential to healthy and agreeable living conditions. Atmospheric pollution from vehicle exhausts, factory chimneys and other fuel combustions causes erosion of ancient buildings and serious respiratory ailments, now the prime cause of morbidity and mortality in several large Indian cities. Disposal of urban waste, garbage, domestic and industrial sewage, contamination of air and water by pathogenic and obnoxious pollutants urgently need systematic monitoring to prevent insult and injury to human health; all need greater planned investments than are presently provided.

A major challenge, often ignored, is to deliver adequate, acceptable and safe food to urban markets and consumers. As cities grow and become more congested, the time to transport food from rural producers to urban processors and consumers inexorably increases. The longer the time taken to transport perishable crops, livestock and fisheries products, greater is the probability of deterioration, spoilage and waste. Pearson, Brandt and Brundtland all described accelerating rates of urban growth and the urgent need for foresight and planning. In several publications this author has advocated systematic analysis of urban-to-rural food distribution, analyses that make use of simulation models to compare alternative routes, times of day or night and modes of transport that provide rural-to-urban food conveyance in an efficient and economic manner. Rural-to-urban migrations could be slowed by providing rural employment opportunities in facilities for primary processing of food and other agribusiness activities [Hulse 1995, 1999].

Table 15 illustrates patterns and predictions of urban growth and rural decline.

Unsustainable urban development

The above data illustrate how urban populations have grown and are expected to grow, while rural populations progressively decline. The implications for food production and security, for maintenance of urban

Table 15
Urban populations

	Total (millions) population				Annual % growth	% of total population			
	1975	1995	2000	2025	1975–2000	1975	1995	2000	2025
World	1540	2600	2900	5100	2.5	38	45	48	61
Asia	590	1200	1400	2700	3.3	25	35	39	55
Africa	104	250	305	800	4.4	25	34	38	54
Europe	445	535	550	600	0.6	67	74	76	83
Latin America	195	360	405	600	2.3	60	74	77	85

	Rural populations: Annual % growth		
	1970–80	1980–90	2000–2010
Developing nations	2.1	0.9	0.3
Africa	2.0	2.0	1.4
Asia	2.1	1.0	0.7
Latin America	0.7	0.03	−0.07

infrastructures, services and transport systems, for efficient safe delivery of foods to urban markets, are gravely disturbing when one recognises how little imaginative attention and investment is devoted to urban planning and management throughout the world. These data and historical records illustrate that urban development is rarely a systematic gradual unfolding that progresses according to a predetermined plan. Unplanned and uncontrolled growth, worsening congestion, decrepit and degenerating infrastructures and services are more typical than exceptional. Urban development, ranging from at best spasmodic and inconsistent to at worst chaotic, cannot by any criteria be considered sustainable.

Rising populations, together with increasing and diversifying demands and resource consumption, exacerbate stress on the earth's life-supporting capacity. Rising urban demands degrade available land, water, air, energy and other critical resources. Despoilment of the environment advances in scale and complexity as urban communities increase in size and density. More reliable methodologies are needed to identify interactions among demographic change, rising demand and stress on resources, services and amenities essential to expanding urban populations [WRI 1997].

Most of the world's over 20 mega-cities (cities with a population of 10 million or more) plus 300 cities each with populations in excess of one million were not designed for, nor are readily adaptable to, the population size and traffic flows now inflicted upon them. Few of the world's large cities are coping satisfactorily with traffic congestion, atmospheric pollution, provision of safe water, hygienic sanitation and waste disposal. Such planning as is evident is frequently *ad hoc*, short term and unsystematic. Systems analysts are urgently needed to design mathematical simulation models to compare alternative systems of transportation, food distribution, delivery of health, sanitation and public transportation services, electrical and other essential energy resources.

The Confederation of British Industry estimates the accrued cost of traffic congestion in Britain in excess of £19 billion, a cost comprised of wasted working time and automobile fuel.

Energy

Antoine Lavoisier cogently commented: "La vie est une fonction chimique": life is a function of chemistry. He had sound reason: every living process depends on the conversion of one form of energy to another by biochemical reactions.

Comparative estimates of energy consumption are complicated by energy being expressed in different units by different agencies. Fossil fuels may be quoted in joules, ergs, calories, British Thermal units (BThU), kilograms or tonnes of oil equivalent, kg or tonnes of coal equivalent, barrels of oil, cubic feet or cubic metres of gas. Electricity is normally measured in kilowatt hours but may be converted to other energy units. Because the calorific value of fossil fuels varies with type, composition and source, among remote and poorer communities energy consumption cannot be accurately measured. All aggregated energy statistics cited by international agencies must be regarded as best indicative estimates. Statistical statements of available reserves are confused by variable states of accessibility and costs of extraction and processing. What price industries and consumers are willing to pay for fossil fuels influences whether the cost of extraction and delivery will be economic to fuel producers and distributors.

The following are approximate equivalents among diverse energy units.

1.0 joule = 10^6 ergs   1.0 BTU = 1054 joules
1.0 calorie = 4.19 joules   1.0 kwatt hour = 3.6 × 10^6 joules

1.0 kg coal = 29.3 × 10^6 joules = 7 × 10^6 calories
1.0 kg oil = 41.9 × 10^6 joules = 10 × 10^6 calories

1.0 kg coal = 0.7 kg oil   1.0 barrel of oil = 0.133 tonnes.
1.0 m^3 natural gas (methane) = 38 × 10^6 joules

[Source: Economist Desk Companion 1992: Pub: *Economist Books* U K]

The following standard abbreviations simplify citation of the large numbers in which aggregates of energy consumption are quoted:

mega (million) = 10^6 giga (billion)   = 10^9
tera (trillion)   = 10^{12} peta (quadrillion) = 10^{15} exa (quintillion) = 10^{18}

Thus a gigajoule = one billion joules; an exajoule = 10^{18} joules

Table 16 quotes World Bank estimates of energy consumption by region during the 1990s, expressed in tonnes of oil equivalent (toe).

Table 16
Energy Consumption 1990s

	Total 1999 Bn toe	1989–99 % incr	per cap 1999 kg oe	Fossil	Nuclear 1999	Hydroelec Bn toe
World	9700	+ 12.7	1600	8000	700	200
O E C D	6000	+ 10.0	4500	5000	600	130
Devg nations	3600	+ 38.5	800	2600	40	90
US/Canada	2500	+ 15.2	8100	2100	200	50
Europe	2600	–	3500	2100	300	60
Asia	2900	+ 43.1	900	2200	100	40
China	1100	+ 29.2	900	3400	900	20
Africa	275	+ 16.4	330	–	–	–
Latin America	530	+ 9.3	800	–	–	–

If all people in all nations consumed energy as extravagantly as Americans and Canadians, total world demand would rise more than five-fold and cause substantial increase in atmospheric pollution. It is hypocritical for North Americans to preach energy and other resource conservation to emerging nations until they themselves become less extravagant.

Forecast future demand

The World Resources Institute calculates that, during the second half of the twentieth century, world energy consumption from all sources more than tripled [WRI 2000]. A publication *International Energy Outlook* predicts that world energy demand from 2001 to 2025 will increase by some 55 per cent, at an average rate of 2.7 per cent a year. Among the fastest growing Asian economies, most notably China and India, energy consumption will double by 2025. Most energy demand will be satisfied in the foreseeable future by oil, natural gas and coal. These predictions assume little real change in fossil fuel prices, a questionable assumption given present political turmoil in the Middle East and production uncertainties among several oil and gas producers.

During 2004, assuming that energy prices would remain relatively stable, predicted annual growth rates until 2025 for natural gas were 2.2 per cent a year, for oil 1.9 per cent, for coal 1.6 per cent, and for nuclear power 0.6 per cent a year. World electricity generation is expected to rise from 13.3 tera kwh to 23.7 tera kwh, the fastest growth – more than 3.5 per cent a year – being among developing nations. Natural gas is forecast to grow from 18 to 25 per cent of fuels used in electric power generation. The assumption that prices will be relatively stable is extremely dubious, given the political disposition of the US government and the aspirations of several major oil and gas producers.

In addition to rapidly rising demands for automobile fuels, energy demand is exacerbated by cheap air travel now increasing at 7 per cent a year. It is to be hoped that, in pursuit of petroleum and diesel fuels, other nations will not follow the unethical example of the US Geological Survey, which in cooperation with US oil companies, proposes to explore and exploit oil and gas in the Arctic Ocean. If, as now seems probable, oil, gas and other fuel prices will continue to rise, non-conventional alternative energy sources could become commercially attractive.

Alternatives to fossil fuels

Biofuels – One might argue that all fossil fuels are biofuels since coal, oil and natural gas originated as trees and other plants that grew many millions of years ago. In contemporary vocabulary, bio-fuels are products of microbial or biochemical conversion of plant carbohydrates or lipids.

Alcohol fuels – Methanol and ethanol are the simplest alcohols used as fuels mixed with petrol for automobile internal combustion engines. Methanol may be produced from natural gas, coal, biomass from woody plants, various wastes and sewage. Ethanol results from fermentation of sugars derived by acid or enzymatic hydrolysis of complex carbohydrates, or by extraction from sugar cane. Used in conventional automobile engines, 5 to 10 per cent ethanol is the maximum that can be mixed with petrol. Higher ethanol, methanol petrol mixtures require specially designed engines, experimental in North America, and known as flexible fuel vehicles (FFVs). Alcohol fuels provide less energy per volume but emit less carbon monoxide and dioxide than petrol. In Brazil, some 60 per cent of motor vehicles use ethanol derived from sugar cane mixed with petrol.

Some years ago, in a private communication, the late Professor John Hawthorn of Strathclyde University calculated that the energy derived from ethanol combustion was less that the total energy expended in planting, cultivating, fertilising, harvesting, isolating, fermenting, distilling and distributing the ethanol produced from carbohydrates. Others have recently expressed similar reservations. Brazil justifies its ethanol fuel programme by the foreign currency saved from reduced oil imports.

Enthusiasm for ethanol derived from edible cereal starch suggests the affluent are more intent on satisfying the appetites of their automobiles than in feeding millions of hungry people in a world where arable land suitable for crop cultivation is fast declining. It would seem more sensitive to human need to derive alcohol fuels from such cellulosic waste as harvested straw, wild grasses and processed timber.

Biodiesel – During the late nineteenth century, Rudolph Diesel, born in Paris of German parents, designed an engine to compete with Nikolaus Otto's internal combustion engine that depends on ignition from a spark plug. Diesel's engine employed high compression to ignite the fuel. Originally he used powdered coal or vegetable oils under compression. Diesel's invention became widely used in many kinds of vehicle and stationary engines, powered by diesel fuel from refined petroleum. There is now interest in fuelling diesel engines with derivatives of vegetable oils.

Biodiesel fuels intended to reduce dependence on petro-diesel are mostly methyl esters of fatty acids produced from vegetable oils: in Europe mainly from sunflower, in the United States from soybeans, in East Asia from oil palm. In several Asian countries expansion of oil palm plantations to provide biodiesel results in extensive depredation of natural forests.

The government of India estimates that the nation's demand for diesel fuel will rise from 60m tonnes to 100m tonnes during the next 25 years. Consequently biodiesel is perceived as an attractive alternative to, or supplement for petro-diesel. India's demand for edible vegetable oils significantly exceeds its production. Indian scientists have therefore identified several non-edible oilseeds suitable for conversion to biodiesel. Particularly attractive are inedible oils from the seeds of *Jatropha curcas.* Jatropha is of special interest, being tolerant of saline and alkaline soils, of near-arid and semi-arid ecologies. Plantations of Jatropha are being established in several Indian states on land ill suited to food crop cultivation. Indians are also exploring biodiesel derived from waste cooking oils, an example which nations addicted to fried foods would be wise to emulate.

Biogas – Produced by microbial fermentation of human and animal excreta and other waste, methane has been used as a household fuel in the People's Republic of China for many generations. A British organisation, Energy Saving Trust, estimates that methane from all human and animal waste in Britain could provide fuel sufficient for 50 per cent of the nations lorries and buses, with a consequent reduction of 95 and 80 per cent of CO_2 and NO_x emissions, respectively, from vehicle exhausts. Sweden captures and uses biogas to fuel many of its buses and refuse trucks.

Hydrogen fuel cells – Engines energised by hydrogen fuel cells emit only water vapour as exhaust. Difficulties arise from hydrogen's being plentiful but not in a free state. Electrolytic separation of hydrogen from water demands large quantities of electricity; hydrogen cannot be compressed or liquefied to equal the caloric density of petrol or diesel. Materials for hydrogen fuel cells are costly, engines that incorporate them are heavy; petrol and alcohols are more easily distributed than hydrogen. Since 1993, several governments and automobile industries have invested substantially in fuel cell research and development. Several authorities suggest that electricity generated from expanded nuclear power could economically energise hydrolytic production of hydrogen.

The principle of the hydrogen fuel cell was first described in 1839, 40 years before Daimler-Benz brought forth their first petrol engine. Fuel cells are based on reverse electrolysis: by electrochemical reaction between hydrogen and oxygen. The cell, composed of two electrodes on either side of an electrolyte, is supplied with hydrogen to react with atmospheric oxygen. At the anode hydrogen atoms converted to ions release electrons as the source of electrical energy. Water is the sole by-product. Proton exchange membrane (PEM) hydrogen fuel cells embody a solid electrolyte and operate at around 80°C. In cooperation with Ballard Power Systems, Daimler-Benz have designed and constructed a bus, big enough for 21 passengers, driven by 24 hydrogen fuel cells with power equivalent to a 160 HP diesel engine. Various vehicles driven by fuel cells are being road-tested in a number of countries, as are electricity generators of differing capacities.

If and how soon hydrogen fuel cells will significantly provide energy to vehicle engines is a matter of conjecture. Their zero carbon emissions are environmentally attractive but the cost of installation and the difficulties of distributing hydrogen to thousands of automobile service stations remain formidable constraints. Nevertheless, given the extensive interest and investment, it seems likely that vehicles powered by hydrogen fuel cells will at some future time be realised.

Some enthusiasts promote methanol fuel cells. As a liquid, methanol is easier to package and distribute than hydrogen. PEM fuel cells operated with methanol are proclaimed as promising replacements for lithium ion batteries. A recent author predicts early advent of laptop computers powered by small methanol fuel cells. Given their environmental advantages and the substantial research investments projected, PEM fuel cells will probably find economic commercial application at some time in the future.

Electric motors – There appears to be growing interest in electric motors for commercial and passenger vehicles. The main obstacle has been the low capacity of storage batteries, limiting the distance a vehicle can travel after being recharged. For short distance travel (up to 100 km) in congested urban roads, electric motors are attractive in that they use no fuel when stopped in a traffic jam and emit no exhaust. For longer journeys, hybrid engines, powered both by gasoline or diesel and electric motors that are recharged by the fossil fuel engine, are under development.

Solar and wind energy – Where sunlight is abundant, conversion of solar energy into electricity, first proposed in 1839 by the French physicist Becquerel, deserves greater attention. Where insolation intensities are high, energy unused from sunlight is equivalent to many times the world's commercial energy. Assuming solar energy conversion efficiency of 15 per cent, the total primary energy consumption of all developing countries could theoretically be provided from receptors occupying barely 0.1 per cent of their total land area, an area less than is now covered by reservoirs for hydroelectric generation. Future costs are speculative but a private communication from an experienced source predicts, between 2010 and 2025, a cost of 5c/kw hour for solar thermal and 10c/kwh for photovoltaic conversion. A recent UN report calculates that, using modern solar conversion technologies, an area of 640,000 km² in the Sahara could generate electricity equivalent to present total world electricity consumption.

Photovoltaic cells convert photons from solar radiations into electricity. Photons are absorbed by semi-conductors such as silicon wafers. Negatively charged electrons are released to produce a flow of electricity. Solar cells are often connected in series, encapsulated in a module covered by a sheet of glass to permit light to pass and protect the semiconductor from rain and contaminants.

Concentrated Solar Power [SCP] systems use mirrors to concentrate energy from the sun onto vessels that contain a liquid or gas that when heated to *ca* 400°C powers steam turbines. SCP has been used to generate electricity in the Californian Mojave desert for more than a decade. Some 30 countries have committed over $13 billion to a nuclear fusion reactor, an unproven and unpredictable power generation technology. A significantly smaller investment in SCP, a technology proven as effective and economic, would be more sensible and productive of clean energy and free of GHG emissions.

Windmills are an eleventh century Persian invention. During the eighteenth and nineteenth centuries, the Netherlands and Germany, respectively, drew power from 8,000 and 18,000 windmills. To generate energy equal to a modern nuclear power station, one experienced authority estimates that the most efficient modern windmills would need to cover more than 350 km² [Barlow 1997]. Portugal appears intent on creating a sizable wind power facility that would generate electricity sufficient to satisfy the demand of 750,000 households. Worldwide, wind power generates close to 60,000 MW. European countries appear most committed. Germany generates 18,500 MW, Spain 10,000 MW, and Denmark 3,200 MW. India reportedly produces *ca* 4,500 MW by wind-turbines.

Nuclear fission and fusion – Many nuclear reactors around the world generate electrical energy by *nuclear fission* and nuclear decay. Nuclear energy is liberated when uranium 235 (^{235}U) is so enriched and concentrated that the radioactive decay in a controlled chain reaction creates heat to produce steam to drive turbines that generate electricity. Nuclear power provides *ca* 7 per cent of total world energy, 17 per cent of the planet's electricity. Roughly 80 per cent of electricity in France, 20 per cent in the United States comes from nuclear reactors.

Nuclear power fell from favour following the 1979 Three Mile Island accident and the catastrophic disaster in 1986 at Chernobyl, when radioactive particles were carried by atmospheric currents as far away as Scotland. Contemporary renewed interest in nuclear energy results from the need to reduce CO_2 emissions and the promise of safer reactors. As evidenced from Chernobyl, humans and other organisms, exposed to intense radiation, suffer severe genetic injury and fatal illness, with gross malformations in unborn living embryos. The earliest reported nuclear fission experiments were carried out in 1938 by German physicists in Berlin. The world's first nuclear power plant was constructed in Obninsk, USSR, during 1954.

Nuclear fusion generates the energy that heats the sun and illuminates many distant stars. Fusion of the nuclei of two hydrogen isotopes, deuterium and tritium, liberate vast quantities of thermal energy. Deuterium can be isolated from water; the heavier isotope, tritium, sparsely available naturally is produced by neutron bombardment of lithium. Nuclear fusion is attractive in its waste products being less hazardous than those from nuclear fission. The obstacles to the economic generation of energy by nuclear fusion are many, most important being that in all reported experiments more energy is required to cause the isotope nuclei to fuse than is generated by their fusion. The largest known experimental nuclear fusion reactor is the Jet at Oxford, England. Several European nations propose a consortium to invest at least 10 billion euros in a massive prototype nuclear fusion reactor. By 2050, it is hoped that nuclear fusion will economically satisfy much the world's energy needs. Sceptics insist that economic and efficient nuclear fusion will not be realised before the end of the century and believe the huge sum would be more productively invested in solar and wind energy.

Tidal power – There is discussion in Britain of a proposal to construct a 16 km barrier across the Severn estuary. The tides being predictable, 175 sluices would be opened to allow incoming tides to fill a huge basin in which water would be held until the tides recede, when the out-flowing water would drive some 200 electric turbines. It is claimed that the power generated would be equivalent to power from two modern nuclear power stations. The predicted cost is $27 billion. Ecologists fear the proposed installation would disrupt many wild life habitats for fish and birds.

Efficiency of utilisation and conservation

How much energy is consumed is influenced by how much is wasted and used inefficiently. Americans and Canadians, accustomed in earlier years to very cheap electricity and other fuels, persist in extravagant consumption by driving large motor vehicles, high power consumption in advertising signs, in public buildings and private homes, over heating during the winter and excessively cool air-conditioning in summer months. The efficiency with which motor vehicles of similar size and engine power burn fuel depends on how effectively they are maintained, how, where and at what speeds they are driven. Energy consumed to heat or air condition a building is determined by design, structure, insulation and patterns of air circulation.

Thermal energy in factory processes is wasted by steam leaking, by poorly insulated steam pipes, by process scheduling that fails to use heat generated economically and conservatively. Prudent managers conduct frequent energy audits of total consumption and use by high energy consuming processes. Large companies can install integrating watt-meters, liquid and gas flow meters to measure actual consumption. Smaller companies rely on historical and account records received from suppliers.

India is believed to lose more than 20 per cent of electric power generated during transmission over long distribution grids. Though the generation cost per unit from small-scale generators is higher than from large central installations, transmission losses from small generators located close to users are significantly lower. To supply power to small rural industries, small generators are more economic and reliable than power from a large electrical grid. Urban motor vehicles when powered by electric motors use no energy when stuck in a traffic jam and emit no obnoxious fumes. Since the mid-1930s in cities in Britain, foods have been economically and hygienically transported in electric vehicles. Electric engines operate efficiently in the higher temperatures of the tropics.

Regrettably, politicians in many nations, unwilling to offend their electorates, lack the courage and commitment to raise energy prices to levels that would encourage greater conservation. North Americans oppose a carbon tax on vehicle fuels that would serve to reduce atmospheric pollution. It is gratifying that the Canadian province of Quebec has recently imposed a carbon tax.

Atmosphere and climate

The earth's atmosphere consists of a thin blanket of air that makes all life possible. One-half of the atmosphere lies within 5.5 km, the outer layers extend to 30 km above the earth's surface. The approximate gaseous composition of pure dry air is as shown below:

	% (per cent)		% (per cent)
Nitrogen	78	Argon	1.0
Oxygen	21	Neon	0.002
Carbon dioxide	0.035 (variable)	Helium	0.0005

Other atmospheric gases include ozone, methane, oxides of nitrogen and various pollutants [Hengeveld 1995].

The globe's climatic system may be regarded as a giant heat engine driven by incoming solar energy. Solar energy warms the earth and surrounding air, motivates atmospheric winds and ocean currents, and drives the evaporation and precipitation processes integral to the water cycle. Solar radiation is the energy source for all biospheric and photochemical reactions. The lowest atmospheric layer, which contains about 80 per cent of the total gaseous mass, is the troposphere. All energy that enters the climatic system eventually returns into space as infrared radiation. If energy leaves at the same rate it enters, the heat engine is in balance and the earth's average surface temperature will stay relatively constant. If the energy exchange becomes unbalanced, average temperatures will change until the system adjusts to a new equilibrium. Nitrogen and oxygen are relatively transparent to radiation and exert little effect on the energy that passes through. The remaining 1 per cent, made up of various aerosols and gases that reflect, absorb, and re-emit significant proportions of incoming and outgoing radiations, is most influential in regulating energy flows that drive climatic processes.

Climate is commonly defined as *average weather*: the average day-to-day variations in temperature, precipitation, cloud formation and winds at specific locations around the globe. Climate is a changeable phenomenon. Together with clouds, minor gases in the atmosphere provide an insulating blanket that keeps most of the earth warm, a phenomenon known as 'the greenhouse effect', caused by 'greenhouse gases' (GHG). Most important GHG are water vapour, carbon dioxide, methane, ozone and oxides of nitrogen. Theoretically, the solar energy that reaches the Planet is sufficient only to maintain an average surface temperature of $-18°C$ but the GHG blanket raises the average surface temperature to around $15°C$. The temperature difference between the tropical zones and the poles is the main driving force for atmospheric winds and ocean currents. Atmospheric moisture is transported by air currents; precipitation patterns are influenced by global atmospheric circulations. Ocean currents, topographic features and rates of evaporation influence the complex, variable precipitation patterns around the world.

The earth's natural climatic system is dynamic and constantly changing. Climate data have been measured and recorded for about 150 years. Information about weather patterns over the past one million years is derived mainly from paleoclimatic indicators. Analyses of Antarctic and Greenland ice-cores indicate a strong correlation between climatic change and atmospheric concentrations of carbon dioxide and methane, two of the main GHG. Global glaciations seem to have appeared at 100,000-year intervals, each being succeeded by a $4°C$ to $6°C$ rise in average temperature. Temperatures in the late nineteenth and early twentieth centuries were cooler than those presently recorded. From the 1920s, average temperatures rose steadily to a peak in the 1940s, declined slightly into the mid-1970s since when warming has resumed, average temperatures reaching new highs in the 1980s and 1990s when the eleven warmest years of the past century were recorded. The present average global temperature is 0.5°C higher than in the nineteenth century. The Planet's average temperature was the highest on record during the first four years of the twenty-first century. Despite denials by the automotive and petroleum industries and the governments they financially influence, human and industrial activities are unquestionably causing severe climate change, the consequences of which could be devastating for many regions.

GHG emissions

It is difficult to determine precisely the quantities of carbon dioxide and methane emitted into the atmosphere, for how long they will remain, and the magnitude of their effect as GHG on climate change. Carbon dioxide is removed by photosynthetic conversion in biotic materials, by direct absorption into oceans, surface waters and plants. CO_2 is released by plant and animal respiration, from decay of biomass, combustion of fossil and other fuels, and volcanic emissions. Giving cause for concern are the increasing quantities of GHG released into the atmosphere, CO_2 concentrations being 20 per cent higher than at any time over the past 200,000 years. Methane concentrations have more than doubled since the late eighteenth century. Table 17 illustrates the state of GHG emissions in 1999 [Data from World Bank, WRI, Environment Canada].

It is not surprising that nations with highest *per cap* consumption of energy emit the highest *per cap* levels of GHG. Of particular concern is the rapid rise in energy consumption, in combustion of fossil fuels and resultant GHG emissions from developing nations, especially those in transition into industrialised economies. Table 17A records estimates of CO_2 emissions by different nations as presented in *HDR 2005*.

Table 17
Greenhouse gas emissions 1999

	CO_2 [Mt]	% change 1990–99	Per cap[t]	Cumulative CO_2 [Mt] 1800–2000	CH4 [Mt]	NOx [Mt]
World	23000	+9	3.9	1.01 M	6300	3600
Asia	6900	+38	2.1	–	2600	1200
PR China	3100	+26	2.5	73000	960	500
Japan	1200	+11	9.1	37000	60	30
Europe	5900	+10	8.1	411000	1200	600
United States	5600	+15	19.9	301000	800	500
Developing nations	8000	+37	1.8	197000	3800	2200

Mt = Million tonnes; t = tonnes

Table 17A
CO_2 emissions 2002

	Tonnes/cap/Yr	% World total
USA	20.1	24.4
Australia	18.3	1.5
Canada	16.5	1.9
Japan	9.4	5.2
United Kingdom	9.2	2.5
Qatar	53.1	0.2
Bahrain	30.6	0.1
Russian Federation	9.9	6.2
P.R China	2.7	12.1
India	1.2	4.7

On average, every automobile in the United States (excluding large trucks and buses) emits 20 tonnes of CO_2 every year. It is estimated that 50 per cent of all the world's exhaust emissions come from vehicles in the United States.

Human activities are disturbing the state and balance of the global carbon cycle. During the twentieth century, massive conversions of forests into agricultural land are estimated to have released over one billion tonnes of carbon. The burning of fossil fuels and other combustibles add some 23 billion tonnes of carbon dioxide to the atmosphere every year, more than 10 times the scale of emissions 100 years ago. Some 50 per cent of carbon emissions may return to the carbon cycle by absorption into oceans, terrestrial ecosystems and forests. Whether these natural carbon sinks can absorb increasing amounts of anthropogenic CO_2 emissions is doubtful given many unknowns: to what size will world populations grow, will future generations depend on present energy sources and will they use energy more conservatively? In the near term it is inevitable that nations such as China and India will increase their fossil fuel consumption with persistent increase in GHG emissions. Nor is there persuasive evidence that those who now consume most fossil fuels and emit most GHG will be converted to considerate conservationists. Without dramatic change, by the year 2050, CO_2 emissions could rise to 400 times present levels.

Methane is released from organic decay, wetlands, rice paddies and the digestive processes of ruminants. Though areas of natural wetlands are decreasing, areas under rice cultivation are expanding. The numbers of ruminants that grew fourfold in the twentieth century are destined to further increase with rising demands for livestock products. Other greenhouse gases, in higher concentrations than a century ago, rise continually. Ozone, naturally present in the troposphere, has increased from various chemical processes to levels 50 per cent higher than in the nineteenth century.

Recent data indicates that since the industrial revolution of the mid-nineteenth century concentrations of GHG have risen by more than 70% and are now increasing by over 2% per year. Each tonne of carbon dioxide emitted inflicts a damage cost of about $85.

Climate and biodiversity

Over the past 30 years, several environmental agencies in Britain have collectively studied climate change, particularly warming attributable to GHG, in relation to shifts in abundance and distribution of biological species. Predictions are derived from modelled projections of future climate change, that associate present and predicted climates including average temperatures, precipitations and seasonal variations with existing species distributions. The authors assess extinction risks among biological species covering 20 per cent of the Earth's terrestrial surface. If the lowest climate-warming projections are realised, 18 per cent could become extinct; with maximum warming projections, 35 per cent of living species could become extinct during the twenty first century. If GHG emissions are not reduced, the earth will become warmer than at any period during past 40 million years causing extensive depletion of biodiversity [Thomas *et al* 2004].

A meeting of the United Nations Environment Programme convened in Curitiba, Brazil, depressingly reports that the destruction of many organisms is occurring almost 1,000 times as fast as happened before heavy industrialisation and motor vehicles became so widespread. In its report *Global Biodiversity Outlook,* the UN lists primary causes of biodiversity degradation as habitat destruction (particularly the stripping of tropical forests), overexploitation of most natural resources, human and industrially induced climate change, all accelerating at rates that are entirely unsustainable. More than 20,000 km^2 of Amazonian forests are destroyed annually by excessive, wasteful logging, by clearing to provide land on which to raise coffee and cultivate soybeans to feed animals destined to supply North American and European fast food stores.

An encouraging recent announcement by Cary Fowler, executive director of the Global Crop Diversity Trust, states: On the Norwegian island of Svalbard, a few hundred kilometres from the North Pole, a consortium of countries plans to establish a super-safe repository "for the world's most precious asset" – over one million samples of agricultural seed. Neither nuclear war, nor global warming, nor any other imaginable threat will disturb the viability of seeds stored inside aluminium packets. The GCDT, based in Rome, will manage the seed storage facility.

International climate programmes

International organisations that study climate change, oceanic and atmospheric cycles include the Geosphere-Biosphere programme that began in 1986, the World Climate Research Programme, the Intergovernmental Panel on Climate Change (IPCC) and most recently the Global Environment Change related to Food Systems (GECAFS). The IPCC states that temperature and precipitation patterns of distribution are determined by topographic features, interactions among oceans, ice and land surfaces and atmospheric circulations. Most influential are:

a. control of GHG
b. effects of clouds and aerosols on radiation balance
c. transport and storage of heat and carbon dioxide by oceans
d. rising sea levels from polar ice melted by global warming.

Though mathematical models present different conclusions to related issues, several that simulate a doubling of carbon dioxide concentrations since the industrial revolution predict that:

a. average global surface temperatures could rise by 4.5°C, with most warming in polar regions;
b. significant changes in patterns of evaporation and precipitation, with the northern sub-tropical zones drier, and higher moisture in polar regions,
c. rising concentrations of GHG will cause climatic change, with serious consequences for ecosystems and agriculture. To avoid catastrophic climatic change nations, industries and communities must be more conservative of energy and other resources.

Scientific evidence from several reliable observers indicates that the Arctic and Antarctic are warming roughly twice as fast as the rest of the Planet and predict a rise in Arctic temperatures of 5–7°C during the twenty-first century. If present trends continue, by 2050 large areas of the Arctic Ocean will be ice-free during summer months. Melting of Arctic ice will cause devastating inundations of many coastal regions. In addition, more than 30 per cent of polar bears and many other Arctic and Antarctic creatures will become extinct.

Intergovernmental Panel on Climate Change [IPCC]

The Fourth Assessment Report compiled in 2006 by the IPCC is cause for serious alarm. It states:

- twelve of the past 13 years were the warmest since records were first assembled
- ocean temperatures have risen to a depth of 3 km below the surface
- snow cover and glaciers are significantly decreasing
- sea levels are rising at *ca* 2 mm/year, and
- by the end of the century, sea levels could rise by 0.5 m, deserts will spread, oceans will rise in acidity causing destruction of coral reefs, and deadly heat waves and violent storms will increase in frequency and severity.

The Stern Report

The British government Treasury Office commissioned Sir Nicholas Stern, former Chief Economist at the World Bank to report on the global economic consequences if present patterns of atmospheric pollution and global climate change persist. The Stern Report is alarming, its findings forebode devastating consequences for future life on earth if present energy generation and utilisation practices persist. At present rates of GHG emissions by 2050 the average global temperature will rise by more than 2°C. Insect-borne diseases, violent storms with flooding of coastal zones, severe regional droughts with acute water shortages will be widely exacerbated. Unless immediately arrested, GHG emissions with consequent climate change will inflict costs equivalent to 20% of global GDP. Remedial prescribed actions taken immediately would cost barely one per cent of GDP per year; resultant benefits would far outweigh the cost of inaction. Failure to act swiftly and decisively will incur far greater cost in the future and will depress global GDP by at least 5%/year.

All countries and ecologies will suffer from failure to implement immediate remedial actions. Present average atmospheric concentrations of CO_2 equivalent are about 430ppm. To realise the desirable maximum level of *ca* 450ppm by 2050 demands CO_2 equivalent reductions of at least 25%. The cost of immediate action is sustainable; failure to act quickly will impose immense hardship and excessive costs on all nations.

Contrary to the myopic pessimism of industrial and political opponents of arresting adverse climate change, creation of more conservative, less polluting energy generation technologies would yield economic benefits for existing energy industries, create new industries and expand employment opportunities in most nations. In addition to energy from renewable resources, more efficient technologies that capture and prevent carbon emissions would permit continued, more economic use of fossil fuels including coal. All industries and communities would gain economic benefit from more conservative use and less waste of energy resources.

Specific actions proposed by the Stern Report include:

a. Encouragement with incentives for all industries and civil society to use energy more efficiently and efficiently with significantly reduced waste.

b. World-wide expansion of emissions trading. Energy-efficient nations should subsidise less-polluting and wasteful energy utilisation among emerging, industrialising nations.

c. Taxation policies that penalise polluters and reward conservationists.

d. International research cooperation is essential to devise and develop efficient, less polluting and resource conserving energy technologies.

e. Significant reduction of deforestation, particularly of natural tropical forests, and extensive programmes of reforestation. Present levels of deforestation result in more polluting emissions than are emitted by all transport vehicles.

f. Effective integration of climate change control in all governmental and international economic and industrial development policies.

Failure among all nations and international agencies to act swiftly and decisively will severely impede survival of future generations, commit the survivors to a miserable existence, and drastically destroy biodiversity by eliminating many species of organisms.

Climate change research

On exceptionally thorough review of recent research on climate and climate change (with 683 references) is available from the Science and Technology Branch of Environment Canada. It reviews GHG, climate during the past 400,000 years, climate model simulation and projections on how to manage the causes and consequences of inevitable climate change (Hengeveld 2006).

Water

Water is the planet's most abundant resource, covering close to 70 per cent of the globe's surface. If all the water were spread evenly across the earth it would form a layer roughly 3,000 metres deep. Almost 97 per cent of the world's water lies in oceans and inland seas, all too saline to be potable or usable for crop irrigation. Of the remaining 3 per cent that is free from salinity, over 80 per cent is sequestered in ice caps and glaciers, exists as atmospheric vapour or is buried in aquifers at various depths below ground (Hulse 1995).

In a sequence of evaporation, transportation and precipitation, water is cycled through the troposphere, each cycle lasting about 10 days. In addition to replenishing lakes, rivers, aquifers, natural and managed ecosystems, water as atmospheric vapour acts as a greenhouse gas. As clouds water contributes to climate control; rain helps atmospheric cleansing by pollutant precipitation. Levels of rainfall vary extensively across the globe, heavy rainfall being characteristic of the Amazon basin, other humid tropical ecologies and the monsoonal regions of Asia. Atmospheric, land and ocean conditions and processes interact to distribute rainfall in geographically uneven patterns, the planet's climatic zones being designated humid, semi-humid, semi-arid and arid. Arid and semi-arid ecologies suffer severe water deficits and cover some 30 per cent of South America, 33 per cent of Europe, 60 per cent of Asia, 85 per cent of Africa, as well as much of Australia and the western plains of North America. More than 80 countries, home to over 40 per cent of the world's population and 30 per cent of cropland, suffer serious periodic drought.

Drought

A discussion of what is known of the history of serious droughts, climatic conditions conducive to drought, the regions and countries where drought is most frequent was published 20 years ago [Hulse and Escott 1986]. Though meteorologists are aware of the general conditions that foreshadow drought, present mathematical models and data available are unable to predict precisely when and where serious droughts will occur. Droughts are often of extended duration, unlike hurricanes and typhoons they do not arrive and depart suddenly. Droughts of varying severity have devastated cultivated crops across the Great Plains of the USA and the Canadian Prairies roughly every 20 years since the 1850s. In the 1930s drought persisted for several years causing egregious crop devastation, soil erosion and poverty for Prairie farmers.

Geophysical evidence indicates that large areas of the African continent suffered long periods of severe drought starting *ca* 10,000 years ago. During the twentieth century serious droughts afflicted the semi-arid territories of sub-Saharan Africa every 15 to 20 years. Particularly disturbing is evidence from West African weather stations that suggest a persistent downward trend in annual precipitations since the 1950s. To what extent drought in Africa will be exacerbated by global warming, aggravated by increasing emissions of GHG, is an issue of contentious debate. Weather conditions over sub-Saharan Africa are influenced by prevailing atmospheric circulation systems, especially interaction between the sub-tropical high-pressure zone and the inter-tropical convergence zone (ICTZ).

As determined by GNP/cap, 20 African countries that are most at risk from recurrent drought are among the world's poorest. For farmers in Australia and North America, some protection is provided by crop insurance and government interventions at times of severe drought devastation. Such programmes of protection are unavailable to most poor African farmers. The total cost to human life and health of African droughts is incalculable. While the plight of thousands with strength sufficient to walk to refugee camps when drought strikes is made known by the world's television and media, many more die unnoticed in their homes or while migrating in search of food and water. Nutritional deficiencies prevail in months immediately before harvest when grain stocks are lowest. Protein-energy malnutrition, anaemia, diarrhoeal diseases, hepatitis and measles, common at all times, are aggravated by drought-induced water deficiencies.

Drought and crops

Cereals and legumes are hygrophytes and need water during their growth cycle. Drought soon after seeding restricts germination; at time of germination drought can desiccate the plantlets before they flower and fruit. Floral development and pollination are highly sensitive to drought-stress which results in a reduced number of seeds. During seed filling drought results in small, shrivelled seeds because of reduced photosynthesis and translocation of carbohydrates from leaves to seed endosperms. Low soil moisture inhibits uptake of soil nutrients through plant root hairs. Plants debilitated by drought are ,susceptible to infection, infestation, invasion by competitive drought-tolerant weeds, and attack by predators. Ravages by locusts following prolonged drought are familiar throughout the African drylands. During the Sahelien drought of the 1970s, sorghum in northern Senegal suffered severe attacks by wild rodents deprived of their natural food sources.

It is sometimes suggested that drought tolerance could be enhanced by genetic modifications. But different plant species use diverse mechanisms to resist drought. Specific adverse effects of drought differ at different stages of growth and development. To imbue a plant with resistance to drought at all stages of growth would be genetically complex. Over the foreseeable future, drought inevitable and unpredictable will inflict severe hardship wherever it strikes.

Water: Available and withdrawn

Table 18, with global and regional data drawn mainly from WRI reports, quotes estimates of available water resources, and water withdrawn in 1995 for various uses.

Table 18
Water Resources 1995

	Available		Withdrawn		Withdrawal %		
	Total (km^3)	Per cap (m^3)	Total (km^3)	Per cap (m^3)	Industry	Domestic	Agriculture
World	41000	7200	3300	645	8	23	69
Africa	4000	5500	145	200	7	5	88
Asia	13200	3800	1630	540	6	9	85
L America	17300	21000	195	410	12	34	54
Europe	6200	8600	455	630	14	55	31
U S A	2500	9400	470	1900	13	45	42

Competition for scarce water

Roughly 70 per cent of the Planet's water and more than 80 per cent in developing countries is used in agriculture, mostly for crop irrigation. Competition for water from industry and purposes other than crop irrigation are progressively increasing. Total world water withdrawn is predicted to rise from *ca* 3,300 km^3 in 1995 to *ca* 4,800 km^3 in 2025, a 50 per cent increase. Over the same period, water for non-irrigation purposes is expected to rise by over 60 per cent, for domestic use by more than 70 per cent, greatest increases being in nations now or once described as 'developing' [Rosegrant *et al* 2002].

By 2025, industrial water consumption will be higher in developing than industrialised nations. Less water available for irrigation forebodes that crop yields among developing nations, now increasing on average by 1.9 per cent a year, will fall to barely 1.2 per cent a year, equivalent to an annual loss of 130 Mt of cereals. While water for irrigation and crop production are adversely affected, world demand for cereals is projected to grow by almost 50 per cent, by some 800 million tonnes. A likely distressing consequence will be the need for poorer nations to raise their food imports from 105 to 240 million tonnes per year by 2025. Most serious shortfall will be in sub-Saharan Africa where imports of cereals are forecast to triple [IFPRI 2002].

Waste and mismanagement

Urgently needed is a worldwide improvement in water use efficiency, in particular for irrigation. Poor irrigation is wasteful of water, depletes groundwater resources and is the cause of salinity and loss of nutrients in soils. To use water more conservatively and efficiently demands technological innovation and administrative reorganisation. In a nation familiar to this author, from a reservoir constructed for hydroelectric power generation, water is gradually released and channelled over long distances for crop irrigation. Between the reservoir and distant farmers' fields the water passes through nine different ministries and agencies of national and state governments. It is not surprising that much water is lost along the way since no single body bears overall responsibility and authority for management and control. When rivers flow through several national territories, administrative difficulties compound and disputes over water rights can lead to violence. Some political analysts anticipate military conflict between Middle Eastern nations who compete for access to limited water supplies. Wasteful misuse is apparent among people who migrate from places with abundant water to arid ecologies. Excessive and extravagant water consumption by foreigners who have taken up residence in new settlements in a Middle Eastern country are deplored by neighbours born in the region and accustomed to use water prudently.

Unlike farmland and buildings, inland lakes and rivers are regarded as common property, free to be plundered or polluted by anyone so disposed. As the worldwide demand for a scarce and precious resource rises, aquifers are being wastefully depleted, surface waters polluted by domestic and municipal effluents, by agricultural and industrial chemical contaminants. Demands for policies that integrate economic with ecological considerations appear to fall on deaf ears.

Future prospects

A contemporary review of world water prospects is published by the International Food Policy Research Institute [Rosegrant *et al* 2002]. The review concludes that the water shortage crisis, predicated on an anticipated worsening of several critical factors, forebodes a breakdown in water supplies for hundred of millions of people, to devastation of wetlands, serious reductions in food production with consequent rise in food prices. Excessive diversions of water flows and overdrafts of ground water will lead to water shortage and related environmental degradations in many parts of the world. In the absence of policies that encourage protection and conservation, competition for water among domestic, farming and industrial consumers will inexorably increase. The review recommends several courses of action:

- investment in infrastructures to increase supply of water for all essential purposes;
- enactment of policies to conserve and enhance efficiency of use;
- investment in research and development designed to raise crop productivity per unit of water and land.

In relation to the last recommendation, it would seem wise, where water is in scarce supply, to encourage farmers to cultivate crops that provide most economic returns to water. In some semi-arid African nations, farmers have switched from subsistence grains to high-value horticultural and floricultural species for export or sale to more affluent communities. Farmers who receive adequate compensation for their produce can be charged realistic prices for the water they use.

Tim Flannery, an Australian climatologist, paints a grim picture of the Planet's water resources. Many of the world's great rivers – the Indus in Pakistan and northern India, the Yellow River in China, the Colorado in the United States and Mexico – are being seriously and irreversibly depleted. Over-draughts of many fossil water sources, accumulated over countless ages, exceed their annual rates of replenishment by aqueous precipitations. The worsening world water crisis is aggravated by climate change and by pollution of inland surface waters. Future water shortages will be most severe in drought-prone regions where already many millions do not have access to adequate safe potable water [Flannery 2006].

A World Bank perspective

In a lengthy policy paper, the World Bank reports how its investments in irrigation, water supply, sanitation and hydro power have encountered operational and social difficulties, water of poor quality for which consumers were unwilling to pay, with poorly managed investments in measures to increase water supply and improve conservation. It admits that the Bank and governments have given too little attention to environmental considerations. The report cites four specific difficulties: a. fragmented public policies and investment programmes; b. failure to recognise interdependence among involved agencies and users; c. neglect of economic pricing and sound financial accounting; d. too little attention to water quality, public health and environmental protection. In its 1993 report the Bank promises future activities will be designed to benefit industries, agriculture, power generation and consumers with cost-control measures commensurate with financial viability.

Water used by industrial biotechnologies

Food processors take in only 4 per cent of the total water consumed by Canadian industries. Nonetheless, all food processors use sizable quantities of clean water: as an ingredient of many foods, as steam, as ice, and for cleaning of raw materials, employees and their clothing, equipment and factory structures, from all of which much waste water emerges. Though governments enact laws to control the quality of industrial effluents, enlightened industrial managers install means to purify and recycle wastewater; and to remove obnoxious or offensive pollutants from all wastewater released [Amos 1997]. Pharmaceutical industries must remove all substances that pose potential hazards to health or ecosystems if released into surface waters and public sewage. Biotechnological industries should purify wastewater to be usable at least for factory cleaning or conversion to steam [Jogdand 1996].

Waste water treatment

Impurities in waste water are categorised as: a. Suspended solids (SS), b. Biological Oxygen Demand (BOD_5), c. Chemical Oxygen Demand (COD); levels of acidity or alkalinity may also be determined. SS covers solid particles large enough to be retained in filters of prescribed porosity. BOD_5 estimates oxygen uptake over 5 days at 20°C to oxidise all biodegradable organic matter. COD determines the quantity of oxygen to oxidise all organic and susceptible minerals. The following table lists typical SS, BOD and COD values for different effluents [Wheatley 1994]:

	BOD (mg/litre)		COD (mg/l)	SS (mg/l)
Domestic sewage	300–600	Meat, poultry	500–8500	75–4500
Silage effluent	30,000 +	Dairy	80–9500	25–4500
Vegetable washing	1600–11,000	Fruits, vegetables	1600–1800	Variable

Jogdand cites BOD values for pharmaceutical, pulp and paper industries' effluents together with a review of biotechnological methods of remediation and purification [Jogdand 1996, 2003]

	BOD (mg/l)		BOD (mg/l)
Sulphite liquor/ paper mills	20000–45000	Antibiotic waste	5000–30000
Distillery stillage	10000–25000	Antibiotic spent liquor	2500–7000

Waste water treatment proceeds in three steps [Wheatley 1994]:

Primary: Physical removal of solids by sedimentation, filtration or activated sludge;

Secondary: Biological degradation of organic matter by aerobic and/or anaerobic microbial fermentation;

Tertiary: Biological and/or chemical inactivation of toxic, organic and inorganic materials resistant to (1) and (2).

Final comment

As it relates to development, 'environment' is virtually impossible to define within precise parameters. As the foregoing illustrates, environmental resources are immensely diverse and difficult to assess accurately. The global environment is complex beyond human comprehension. Recommendations anent global environmental conservation can be prescribed only in broad generalisations since conditions, resources, opportunities and constraints vary extensively among regions, ecologies and communities. There being no international organisation with authority and power to enact and enforce regulations deemed desirable to protect and provide for future generations, recommendations can be little more than hoped for good intentions. Among governments, organisations and individuals, opinions differ widely on desirable courses of action, some differences being honest and intellectual, others motivated by egocentric avarice with no wish to change a wasteful way of life.

It would appear that to many automobile producers, the politicians and governments they support, influence and control, motor vehicles and their appetites are of far higher priority than the needs of millions of impoverished humanity for food, clean water and healthy lives. A statement by a recently retired senior executive of one of the world's largest petroleum corporations, while he received millions of dollars in retirement benefits, that there is no scientific evidence for global warming caused by GHG, was repugnantly greedy and utterly irresponsible.

More than a decade of contentious negotiations resulted, as of April 2006, in 163 countries signing the Kyoto accord that proposes substantial reductions in atmospheric pollution. It is encouraging that, despite the refusal of the US Republican government to ratify the Kyoto accord, the State of California and close to 200 American cities, with combined populations in excess of 100 million, have adopted climate conservation protocols similar to Kyoto accord prescriptions.

Among persons and agencies with genuine serious concerns there seems broad agreement on the need to protect the atmosphere from pollution and human activities conducive to climatic change; to protect fragile ecosystems and conserve biodiversity. The enlightened agree also on the need for informed planning of land use, urban growth, water and energy utilisation and conservation; for equitable opportunity in international trade and access to essential resources; and for the affluent to be more generous in support of the less fortunate.

That being said, these are all issues remarked upon by the Pearson, Brandt and Brundtland Commissions, by countless subsequent international conferences and working groups. The tangible responses to these issues have been far from what the originators hoped for. What should be specific objectives for sustainable environmental development remain contentious and elusive and will probably remain so during the immediate future. The greatest hope lies in exemplary policies and actions by particular governments and communities, policies and actions demonstrably beneficial that others could adapt according to their needs, opportunities and resources.

The costs, insults and injuries to the healthy lives of future generations caused by present extravagant use and waste of energy, water and other essential resources, together with pollution of the atmosphere, cannot be predicted. Many serious observers, most recently Sir Nicholas Stern, believe the consequences will be catastrophic. It is therefore incumbent on all governments, societies and industries to assume the worst, to adopt practices and life styles more conservative and protective of the environment and the planet's resources.

Little will be gained by convening more international conferences. Urgently needed is an international organisation recognised by all nations as having the authority to prescribe standards and regulations for biodiversity and natural resource conservation, for atmospheric protection, to restrict severely and reverse global warming caused by human and industrial activity. Since such regulatory control would demand a less wasteful and politically unpopular life-style, it is unlikely to be proposed or adopted in the foreseeable future. Regrettably, the insatiable greed of the most affluent and powerful nations does not inspire confidence that Brundtland's objective of safeguarding the needs of future generations will come to fruition.

Marie Antoinette is reported to have said: "Let them eat cake". Contemporary affluent and powerful politicians and those who sustain them seem to say: "We can be as wasteful as we please; let future generations suffer and, if need be, suffer death by starvation".

Al Gore, a former US Vice-president, in a pamphlet accompanying his film *An Inconvenient Truth*, lists "ten things to do to help stop global warming". Among them are the following: drive less (every two miles you don't drive saves 1 kg of CO_2); recycle waste; use less hot water; plant trees (a single tree will absorb one tonne of CO_2 over its lifetime); avoid products with a lot of packaging (you can save some 500 kg of CO_2 a year if you cut down your garbage by 10 per cent). These are simple things to do, if the affluent would only take heed.

It is clearly evident, given the political dispositions of American, other affluent governments and the industries that support and encourage them in their wasteful depletion of scarce resources, now evident among emerging industrialising nations, that there is little governmental concern for the needs of future generations, many of whom with economic and environmental conditions worsened by global climate change will be condemned to lives of extreme poverty, hardship and misery.

The US and Canadian governments, who until recently refused to admit that global warming caused by GHG emissions was a scientific fact, now reluctantly claim to be partial believers. It is too soon to know whether they will enact legislation that will offend their industrial supporters in the fuel oil and automotive industries. In any event anything less than drastic action will be too late to reverse the damage done and inevitable future devastations.

11

Case Studies of Successful Projects

In the previous ten chapters the author has taken a critical look at what is often termed 'sustainable development' and has argued continuously for a clear definition of terms of both those words, not only as concepts but also in their application to many different circumstances, from food production and food security to industrial biotechnologies. Too often terms are ill-defined and consequently the objectives are open to misunderstanding. Too often the research is focused on a first, part-way objective – for example, the breeding of higher yielding rice genotypes – and the attainment of the real objective, the adoption of new technologies by small farmers, is either unplanned or else taken for granted. This book has laid heavy emphasis on the importance of a systems approach that rejects any top-down practices and involves the intelligent participation of the beneficiaries from the outset.

This chapter moves from setting out such argumentation to offering a number of examples of where development has been planned on a systems basis, and where the outcome has been sustainable and beneficial. The examples are drawn from Latin America, Africa and Asia, and the International Development Research Centre has been supportive in all but one of them. The exception is the Canada-Mysore Project, where men and women from 45 nations have had training in food technologies; it was launched a decade before IDRC's own birth. The first example of IDRC initiatives concerns its advocacy of farming systems research, to take results from the experimental plots of the international (or national) research institutes to sustainable application in ordinary farmers' fields.

Closely connected is the work done in the creation of a new generation, beyond IRRI and CIMMYT, of international agricultural research institutes: ICRISAT and ICARDA, to focus on research in the semi-arid tropics and other dry areas. With the same motivation, of applying a systems approach to forestry for the purpose of lasting human benefit and resource management, IDRC first launched a social forestry programme and later helped establish ICRAF, the International Centre for Research in Agroforestry, in Nairobi. There is a similar thread of resource management in the fifth case-study, "Local management of water", in which management of demand is particularly stressed, as well as (significantly in China) the involving of the local people affected in selecting the technical solutions.

Once again, the systems approach, starting with a thorough diagnosis of a particular situation by a trained local team, is applied in the case of "Essential Health in Tanzania". And finally, the example in the Pondicherry region of linking a rural population by radio and computer to a wide range of relevant information is an indicator of how imagination and targeted aid with some modern technologies can transform communities.

The Canada-Mysore project

In 1960, the British government convened a conference in London to mark the centenary of the world's first food and drug act, enacted in 1860. During the conference, Joseph Hulse, then research director of one of Canada's largest agribusiness companies, was asked by an old friend, Sir Norman Wright, who was deputy director-general of the UN Food and Agriculture Organisation (FAO), if Canadians would be willing to support FAOs recently launched Freedom From Hunger Campaign. Specifically, Wright asked for financial support for an International Food Technology Training Centre. The Indian government had promised physical facilities for such a centre in a former palace of the Maharajah of Mysore.

An FAO study had revealed that no university, college or other institution anywhere in Asia was providing training in practical means of food preservation and its safe distribution. Countless tonnes of perishable foods were being spoiled and wasted after harvest. Money was needed to cover travel, accommodation and related administrative costs of bringing men and women to the Indian government's Central Food Technology Research Institute (CFTRI) in Mysore, southern India. CFTRI employed a staff of scientists, most of them with degrees in natural and biological sciences from foreign universities. These Indians, together with invited lecturers, would provide training in food preservation technologies to Indians and other Asians.

Early in 1960, Canada had established a small Freedom From Hunger Campaign (FFHC) committee chaired by Mitchell Sharp, a former senior public servant who later served as Foreign Minister in the Liberal government of Pierre Trudeau. The FFHC committee agreed to undertake what became known as the Canada-Mysore Project as its principal contribution to the FAO campaign and appointed Hulse as the chair and director of the project. During a visit to India, Hulse took and collected many photographs illustrative of vast quantities of harvested crops and livestock products that were degraded and spoiled for want of effective preservation. Adrienne Clarkson, then a CBC producer, was the first journalist to interview him on this project. Subsequently, sets of photographic slides with accompanying explanatory notes were provided to Canadian scientists and others, who gave talks to service clubs, schools, religious and other organisations to gather support for the project.

Relatively soon, the presidents of Canada's 20 largest agribusiness corporations formed an industrial fundraising committee, while the

United Nations' Association in Canada, Oxfam, Save the Children Fund and several other NGOs agreed to support the project. The prime minister and the leaders of all parliamentary parties became official patrons; the army, air and sea cadets and many high schools raised money by selling ballpoint pens "To write-off hunger". Local fundraising committees established across the country gave valuable support.

The training programme at Mysore consisted of a series of short courses and a longer, two-year course. The first short course, organised by two Canadians, dealt with storage and de-infestation of cereal and legume grains. During a later visit, Hulse obtained the consent of the University of Mysore to award M.Sc. degrees to all who successfully completed the two-year course. Every major agribusiness company in Canada provided the money for one or more M.Sc. students, as did each of the four Canadian Banks. This was the first time that Canadian banks and most industries had contributed to a charitable project outside of Canada. Canada's two most loved entertainers, John Wayne and Frank Shuster, became honorary co-chairs and recorded 30-second advertising spots that were broadcast across Canada.

The fundraising began in 1961. By 1964 the first short courses began; in 1965 the first batch of 20 M.Sc. students arrived, each from a different Asian nation. Monies raised in Canada supported the IFTTC students during the centre's first seven years of activity. Later, other nations contributed, the Swiss began a school for milling and baking and, in the 1970s, IFTTC became a campus of the United Nations University.

When the project began, India was receiving substantial quantities of Canadian wheat and other food resources in the form of food aid. From the sale of this food, the Indians acquired resources that were invested to build a hostel where the visiting students were housed and fed.

The training programme still goes on. Since it started, more than 7,500 men and women from some 45 developing nations have taken courses at Mysore. Though most were Asians, students from Africa, the Middle East and a few from Latin America have benefited from various training programmes. They have come from as far abroad as Mexico, the Sudan and Zambia. Several subsequently advanced to become directors of food research in Korea, Singapore, Sri Lanka, India and some other nations. Other Asian nations have since created relevant training facilities, several with staff trained at Mysore.

The IFTTC is now self-supporting with its own administrative staff. The courses are so popular that there are 20 to 30 applicants for every place in the two-year M.Sc. course. More than 90 per cent of the graduates are employed in industry. It might be argued that India has benefited most from the Canada-Mysore project, but trainees and graduates from many other nations have returned to teach or work in food industries. Illustrative of the overall benefit: food processing and distributing are the fastest growing industries in several Asian nations, not least India where in 2005 the value of industrially processed foods was over 1,000 times the value when the Canada-Mysore Project started in 1961.

The project has been successful because

a. the objectives, resources available and needed were clearly defined;

b. Canadians created an effective organisation to support and monitor the project;

c. the Indian government, the scientists and technologists at Mysore have maintained a sustained dedication;

d. the Indian food and agribusiness industries have recognised the value of the men and women trained and employed them, and have provided financial and technical support for the Centre;

e. Hulse, in continuing as Canadian Chair and Director of the Canada-Mysore Project, has maintained an active interest in the Mysore Centre and its progress over the years, and during the past 12 years has acted as Visiting Professor, lecturing in agribusiness management and food systems analysis and integration.

Farming systems research – IDRC initiatives

The IDRC Board set the Centre's agricultural staff a special challenge when it determined to place high priority on marginal and impoverished lands and ecologies. For ecologies (then and now) differed widely, as did land use and water management; and the basic infrastructure for these lands and access to markets were scarcely developed. Scientists had done little work on how small farmers managed to eke out a livelihood from any particular quality of land. There was, in fact, in 1970 very weak scientific underpinning on which to base any development help to farmers.

Three decades have seen great change with the worldwide acceptance of farming systems research (FSR). And much of the change and the evolution of FSR can be traced to IDRC's early initiative and consistent support.

Latin America was a logical region in which to start, since researchers in Colombia and Mexico had turned to studying the realities of small-scale farmers who were interplanting crops of corn, beans and potatoes. But about the same time, IDRC president David Hopper, who had worked with the Rockefeller Foundation in India, learnt that an American soils scientist had begun a 'farming systems' project at the International Rice Research Institute (IRRI) in the Philippines, and asked Joseph Hulse, director of IDRC's Agriculture Food and Nutrition Sciences division (AFNS) to see if it warranted the Centre's support.

Hulse found that the project was largely confined to a soil and water study related to rice cultivation, but IDRC agreed to help by financing two soil scientists to produce a soils map of Asian rice-producing territories. As well, through former colleagues in the FAO Land and Water Division, Hulse learnt that FAO had an immense collection of water resource studies in Asia, which had never been systematically collated and analysed. He recruited a post-doctoral fellow, who spent weeks at FAO sifting through the studies and compiled a summary report that was given to IRRI.

Early in the 1970s the Rockefeller Foundation transferred the agronomist Richard Harwood to IRRI to take charge of FSR, but he soon found that the IRRI Board was opposed to any activity other than the breeding of higher-yielding rice genotypes. Fortuitously, its Director-General resigned, and the Board appointed Nyle Brady, Dean of Agriculture at Cornell University, as the next Director General.

Hulse seized the opportunity and visited Brady at Cornell on three occasions to urge that IRRI include a significant FSR component in its programme.

He drew the analogy with marketing research (having been Director of Research at the Canadian agribusiness Maple Leaf Mills). Before launching a new product, food companies identify specific consumers and consult with them about the quality, quantity and price likely to be accepted. Hulse suggested that IRRI researchers likewise determine the capacity of Asian rice farmers to adapt to and adopt new rice production technologies. Brady expressed serious interest, and Hulse promised to seek IDRC support for a broader programme.

The Caqueza project

Meanwhile, IDRC had decided to support the work of the Colombian Farming Institute (ICA), which was starting to study the production methods of farmers in the mountainous region of Eastern Cundimarca. The focus was on nine municipalities centred on Caqueza, and IDRC contributed the skills of Dutch-born agronomist Hubert Zandstra.

Much of the early ICA work there was demonstrating improved methods, rather than researching the local realities and diagnosing the constraints. But it shifted during Zandstra's four years to addressing the factors that may influence farmers in (or deter them from) adopting new methods. He listed them in the book *Caqueza: Living rural development* he co-authored: risk, credit (perhaps rotating funds), marketing, training and buffer institutions (including time for reflection). One main lesson from Caqueza was that the small-scale farmers must have meaningful participation in the research, including sharing the setting of the agenda.

In 1973 another IDRC scientist Gordon Banta visited Caqueza from his base at IRRI in the Philippines. He had been sent to IRRI by David Hopper as a social scientist to help broaden their research for small farmers beyond work on improving crop yields on IRRI's own land. Banta had worked with Richard Harwood on diagnostic studies and field trials on farmers' fields in Batangas province. Banta carried back the Caqueza experience, and in 1974 he and Harwood organised a meeting at which Asian, Latin American and African researchers shared and synthesised their experiences of the past decade. Out of this meeting grew the Asian Cropping Systems Network, which IDRC supported for many years.

Until this network began to influence attitudes, many national institutions did not allow their researchers to conduct their work in farmers' fields. Research managers saw little need to involve, from the outset of introducing new technologies, the farmers affected by it. Some indeed believed that soil and environmental conditions faced by small-scale farmers were similar to those on experimental farms. It took years to change these top-down attitudes in some countries. The on-farm research in the 1970s involved some farmer participation, but there was little methodology developed.

Ilo-Ilo and diversifying FSR

When Harwood resigned to return to the United States, Zandstra moved to IRRI for five years and helped face this lack, by leading an expanded and diversified FSR programme. Much experimenting with methods of systems research took place. A major innovation was to consider as variable factors in land use not only such matters as soil, weather and economic and social conditions, but actually consider land use itself a variable. Did land that had always been planted to rain-fed rice always have to be so? With a 90-day growing season for new high-yield strains, a farmer could also grow cowpeas or soybeans or a second rice crop on the same land – or maybe grow some cash crop and no rice at all. It took time to persuade top staff at IRRI that rice was not a "given" in this method, or those at CIMMYT who were moving into African research that maize was not. IDRC supported experiments of this sort at three sites in the Philippines including a base in Ilo-Ilo, a distance away from IRRI at Los Banos.

Farming systems researchers explored in several other directions. The FSR programmes combined concepts differently and produced many variations in types of farmer participation, the importance placed on diagnostic methods, the market-based analyses, and the dynamics of rural families, which began to give importance to gender analysis. Work done at Ilo-Ilo alongside IRRI produced not only a general broadening of minds about land use; it also led to the formation of a group of female professionals who brought gender aspects to the forefront. It also supported in various countries the establishment of Farmer Field Schools and local agricultural research committees, bringing in not simply farmers but also people at various points along the food and market chain.

On-farm research in land use during the 1980s helped increase productivity in many crops – corn and potatoes, legumes and rice, and also crops and livestock. It is fair to claim that these developments would not have happened if research had only been carried out in single-crop studies or individual cattle-farming enterprises. The FSR approach became adopted by most national centres, as well as in newly created international centres in agroforestry and livestock research (both based in Africa) as part of their official programmes.

The Brundtland report and the Rio conference's Agenda 21 added the challenge of taking into fuller account in the FSR approach the environmental aspects and the sustaining of natural resources. Researchers now had to measure existing and alternative resource uses against these factors and deal with complex trade-offs in land use.

Zandstra, who in the 1990s moved on to head the international potato centre (CIP) in Lima, Peru, made it a priority to build the capacity to manage the natural resources in an unquestionably sustainable way, and in this FSR was an ally. He says: "It is evident and encouraging that FSR has evolved to a set of effective participatory research approaches". And he gives credit to the Ottawa centre: "It is still easy to recognise the impact IDRC has had on the scientists and their institutions through three decades of support for FSR. It has done so by being consistent and by finding important allies in the donor community".

Comment from M.S. Swaminathan

Professor M.S. Swaminathan, former Director General of the International Rice Research Institute, makes the following comments on the subject of the Farming Systems project:

IDRC's visionary programme of Farming Systems Research opened up new opportunities for work, income and ecological security for rice farming families in Asia. The IDRC support of FSR led to a synergy between ecology and economics, and provided work and income security throughout the year. It helped to maximise the benefits of cubic volume of air, soil and water. It showed how we can link paddy and prosperity together by adopting appropriate crop and livestock in efficient integrated farming systems.

Farming systems research spearheaded by IDRC helps to achieve a paradigm shift from the Green [Revolution] to an Ever-Green Revolution designed to enhance productivity in perpetuity without associated ecological harm. During the period I was Director General of IRRI, we observed that the farming systems research led to the integration of efficient stem modulating nitrogen-fixing species. Also FSR helped to develop the international fertiliser response network into an international sustainable rice farming network.

During the twenty-first anniversary of IRRI in 1985, outstanding farming women and men who participated in a special discussion all mentioned that the FSR has opened a new chapter in the art and science of rice cultivation. It was therefore an act of great vision on the part of Dr J.H. Hulse and IDRC who initiated these important research programmes far in advance of the [general] realisation that rice cultivation will have a future only if the principles of ecology, economics, gender and social equity, and income generation are integrated in the form of a farming systems research programme.

Employment in rural agribusiness

Rural people can be liberated from poverty when enabled to earn a stable income from sustained employment in agribusiness enterprises which, individually or collectively, they own and operate. Agribusiness includes production of crops and livestock; of materials essential to crop and livestock production; preservation and processing of agricultural, horticultural and livestock products; and safe delivery to identified markets. Sustained development of rural agribusiness requires:

a.   a comprehensive, critical assessment of resources available and resources needed to produce and deliver familiar, acceptable, affordable products to identified accessible markets;

b.   training to acquire relevant practical knowledge and skills;

c.   constant, consistent, reliable access to technical, financial and marketing advisory services;

d.   organisation into legally registered community cooperative associations administered by the community. A community cooperative organisation can exert greater influence than isolated individuals in gaining favourable access to loans, credit, bulk purchasing and markets; in providing essential supporting and advisory services;

e.   two highly successful programmes in south India are exemplary.

The MSSRF Biovillages programme

In 1991, the M. S. Swaminathan Research Foundation (MSSRF), based in Chennai, began its Biovillages programme for rural agribusiness development. The programme persuasively illustrates that, given the opportunity, with effective training and advisory services, poor rural women with little formal education can collectively and individually engage in sustainable agribusiness. MSSRF supports a national biovillage network, illustrated by the programme in Pondicherry. The Pondicherry programme is administered by a Biovillage Council of some 30 women representatives from 19 rural communities. A Biocentre is staffed by young trained advisers recruited from the biovillage communities who train, assist and provide an immediately accessible advisory service to the rural women, negotiating low interest loans and credit, providing veterinary and ancillary advisory services. Each village community is equipped with a computer linked to market outlets to provide instant access to current prices and levels of demand for products offered by the biovillages. Computer graphics and illustrations are useful tools for demonstration and training.

Pondicherry biovillage women cultivate horticultural crops, produce low-cost hybrid seed for more than a dozen vegetable species, organic fertilisers from vermiform compost, and bio-pesticides from Neem trees and other local resources. The women raise poultry, culture fresh water carp in village ponds, compound fodder for their cows. The Biovillage Centre maintains demonstration plots to ensure efficient crop cultivation and crop combinations that contain most nutrients essential to human health. The Biovillage Women's Council study opportunities for new and expanded activities, manage and monitor collective financial resources, and make bulk purchases at wholesale prices of essential supplies.

VGKK/KT-CFTRI health and employment programme

The Vivekananda Girijana Kalyana Kendra/Karuna Trust is an NGO registered in India that provides health care, basic education, vocational training, biodiversity conservation, stimulates community self-help and small-scale rural industries in the B. R. Hills/Yelandur taluks of Chamarajanagar district of Karnataka State. The forest reserve covers more than 350 square kilometres and is inhabited by some 80,000 people of the Soligas tribe whose ancestors have subsisted on the products of the forest for countless generations. Several years ago an eminent Indian physician abandoned his practice in Bangalore to provide much needed health care to the poor tribal forest people. At the outset, more than half the children suffered chronic malnutrition; infant mortality exceeded 150 per 1000 live births; leprosy, epilepsy, tuberculosis and diarrhoeal diseases were hyper-endemic. Assisted by other volunteer physicians and surgeons, the director provided a diversity of remedial and health care services. They established clinics for maternity, ante- and post- natal care, a surgery, ward residence facilities, a mobile surgery to visit remote communities, and trained young tribal women as nurses, midwives and medical assistants. By multi-antibiotic drugs and reconstructive surgery, the incidence of leprosy dropped from over 20 per cent to less than one per cent of the population. Comparable improvements occurred with other hyper-endemic diseases.

The need was evident for education, vocational training and employment to alleviate abject chronic poverty. Schools for primary, secondary education and vocational training were established. The physician director then requested assistance to promote rural agribusiness industries to harvest and process non-timber forest products (NTFP) and horticultural crops cultivated in forest clearings. Extracts of more than 60 identified natural medicinal plants have been registered as Ayurvedic drugs. Opportunities by which to expand from domestic to small commercial scale processing of fruits, vegetables, wild honey and medicinal plants from the forest were realised through a grant from a development agency and a loan from the local agricultural bank. Roughly 50 poor women were trained to operate and manage factory-scale processing technologies by which to transform various NTFP into marketable products, products familiar and readily acceptable to local people. The factory, with a working area of 220 square metres, is equipped for steam processing, fluid concentration, dehydration and simple fermentations. An ingenious evaporative cooling chamber preserves fresh and perishable biological materials.

The factory, constructed entirely of local materials by local labour, was completed in 45 days and began full-scale production in September 2004. Technologists from the Indian Central Food Technological Research Institute (CFTRI) at Mysore trained the women, gave guidance in product design, development and quality control, and continue as an accessible, reliable source of technical and marketing advice. The

VGKK organisation is formally registered as a cooperative association and administers a credit union from which members can obtain loans or make interest-bearing deposits. In addition to the 50 women who manage and operate the factory, cultivation and harvesting of raw materials, marketing of processed products, provides employment for very many more. Among the tribal people, it is a self-imposed discipline that quantities of NTFP harvested never exceed the natural rate of regeneration. Thus it is a model of biodiversity conservation.

These two examples clearly illustrate that poor rural people who have little formal education can be trained to produce and process a wide diversity of biological materials, so long as they have constant and consistent access to technical, financial and marketing advisory services, services provided by intelligent young people indigenous to and therefore familiar with the language, culture and customs of the people to be assisted. They illustrate that rural poverty alleviation demands very many relatively small financial investments; that poverty alleviation is a labour-intensive not a capital-intensive undertaking, a lesson yet to be learned by many development agencies.

Focus on semi-arid tropics

During its first meeting in October 1970, the IDRC Board of Governors recommended that the Centre's programme give priority to rural poor. Joseph Hulse, then the director of the Agriculture, Food and Nutrition Sciences division (AFNS) proposed, given the Centre's limited resources and the many millions of rural poor, that AFNS concentrate on the poorest of the poor in the semi-arid tropics (SAT). This was agreed and AFNS recruited Hugh Doggett, one of the world's leading sorghum breeders who had spent more than 25 years working in the SAT of Africa. From the outset AFNS pursued a total systems principle in defining its programme and projects. Consequently, over subsequent years AFNS devoted significant resources to both agricultural production and post-production systems.

Production systems

AFNS encouraged and supported production systems research in sorghum, various millets, grain legumes and oil seeds in north, west, east, southern and central Africa, and in the Middle East and Asia. Its support served to strengthen national sorghum and millet improvement programmes throughout francophone West Africa, and provided the first foreign aid grant to the newly independent Zimbabwe in 1980–1. The station at Matopos in southern Zimbabwe later became a regional sorghum and millet improvement centre.

IDRC was a founding member of the Consultative Group on International Agricultural Research (CGIAR) and AFNS acted on behalf of the CGIAR in creating the International Crops Research Institute for the Semi-Arid Tropics (ICRISAT). Since 1971 ICRISAT has maintained headquarters near to Hyderabad, Andhra Pradesh, and a diversifying programme throughout the world, mainly devoted to the improvements of sorghum, millets and groundnuts.

AFNS provided a team of four that recommended an integrated programme for Africa, which would have functioned from four strengthened national centres in Senegal, Nigeria, Uganda and Tanzania. Regrettably, the AFNS team's advice was not accepted and, eventually a new institute was constructed by ICRISAT in Niger.

AFNS' production research for crop improvement in the SAT sought always to strengthen indigenous capability, by linking universities and nations' agricultural programmes with ICRISAT, in addition to creating cooperative networks among several national SAT programmes. To this end, AFNS encouraged and supported a grain legumes cooperative programme among poorer nations located around the Mediterranean littoral. As a result, in 1975, CGIAR invited AFNS to act as executing agent in designing and creating an International Centre for Agricultural Research in Dry Areas (ICARDA).

It was originally intended that ICARDA undertake and coordinate research from three stations: (a) in the Bekaa Valley of Lebanon, (b) in the highlands of Iran, (c) close to the Euphrates in Syria. In consequence of various civil disturbances and changes in government, it was necessary to locate the headquarters and main station at Aleppo in Syria, with a highlands station in Turkey. Later, research resumed in the Bekaa valley. ICARDA continues to provide research support to cereal and legume producers throughout the drylands of West Asia, the Middle East and North Africa. Sorghums and millets were also developed in the Middle East as forage crops.

Sorghums originated in Ethiopia. To expand the range of germ-plasm available to sorghum breeders, AFNS financed sorghum germplasm collection missions to Ethiopia. Similarly, in cooperation with ICARDA, AFNS supported collection of chickpea in several West Asian territories from which chickpea is believed to have originated. By these means, AFNS enabled plant breeders to diversify the agronomic, functional and nutritional properties of these SAT crops.

It was AFNS policy to provide foreign 'expert' assistance only when essential resources did not exist within the SAT nations. A particular difficulty with certain genotypes of sorghum is the anti-nutrient polyphenols (tannins) biosynthesised in the pericarp and/or testa, the outer seed coats, of the grain. Many reported related studies assumed the sorghum polyphenols were hydrolysable tannins, similar to the substances extracted from oak galls (the ancient Celtic and Breton names for the oak were, respectively 'tannum' and 'tann') and other biological sources used since ancient Babylon in the tanning of animal skins to produce leather. AFNS identified one of the world's most experienced polyphenol chemists, Professor Edward Haslam of Sheffield University, provided him with a post-doctoral fellow, Dr Raj K. Gupta, who in due course demonstrated that sorghum polyphenols were condensed tannins related to many biosynthesised in the fruits and seeds of woody plants. This revelation made possible a more comprehensive understanding of sorghum tannins, their concentrations, functions and means of reducing their adverse anti-nutrient effects in breeding programmes.

Post-production systems

It is essential in devising genotypes and phenotypes that generate higher yields (weight of edible material per ha/per yr) that the crops harvested be acceptable and usable by consumers for whom they are intended. From an extensive study in francophone West Africa, AFNS learned that many rural women devoted up to five hours every day to laborious pounding of sorghum, millets and dry legume grains into flour or meal. AFNS undertook an examination of grains structures to discover how the outer seed coats (husk and bran) could be mechanically separated from the endosperm, which when pulverised provides usable flour. Eventually, in cooperation with a farmers' cooperative in Maidugri, northern Nigeria, and Canadian agricultural engineers in Saskatchewan, a simple and effective milling machine was developed and installed in African rural mills.

The mill used rotating carborundum disks to remove the bran, air currents carried the bran out of the mill to be used as animal feed, the dehulled endosperm was collected and reduced to flour of various granularities in a hammer mill. A large group of young Nigerian women determined the properties of sorghum, millet and legume flour most acceptable to various consumer groups, the results indicating the quality of milled products most likely to be purchased. It was a firm AFNS principle that all post-production research be preceded by a consumer survey, to determine what properties were critical to acceptance and utility.

The sorghum milling activity became known to other African communities, with the result that AFNS supported small-scale rural mills in several nations. Carborundum tends to wear and to fracture when rotated above 1000 rpm, therefore, a later development replaced carborundum with lighter, thinner thermoplastic disks that could be rotated at up to 6000 rpm, thus increasing the durability and efficiency of the mills. For some 20 years, mills so constructed have been manufactured in several African countries. A study by a rural sociologist in Botswana discovered that their relief from many hours of pounding had enabled rural women to devote the time liberated to child care, the tending of kitchen gardens and the raising of poultry.

During the early 1970s, several donor agencies tried to persuade poor African farmers to construct small grain silos. The inherent weakness of all the proposals was that they relied entirely on materials to be imported from the donor country. AFNS encouraged a study of existing grain silos, constructed of locally available materials. Applying the recognised principles of harvesting and threshing before sunrise, when diurnal temperatures were lowest, drying by use of sun and prevailing winds, insulation with mud and protection of the silos from direct sunlight, grain sufficient for the annual needs of a typical rural West African farm family could be protectively stored in silos that cost only the local labour to construct them.

Sustaining R & D in SAT

It can be stated with some confidence that the AFNS programme in the SAT, started in the early 1970s, has been sustainable. ICRISAT and ICARDA continue to provide research and training support to many SAT nations, small grain mills continue to be constructed and to function in several African nations. Where rural programmes have collapsed or been debilitated, as in Zimbabwe, has largely resulted from government mismanagement and/or from civil or inter-communal strife.

The integrated systems principle

The AFNS concept of integrated systems, to link production effectively with post-production, is now accepted among many organisations, not least among some of the world's largest transnational corporations who have created a Sustainable Agriculture Initiative, in which production is as much concerned with useful and nutritional properties as with high yield. Primary processing of perishable crops in rural areas, close to sites of production, serves to reduce post-production losses, a matter of considerable importance as cities become more congested and the time to transport harvested crops from rural areas to urban factories inexorably increases.

Social forestry and agroforestry

From past experience, several IDRC staff were persuaded that progress in agricultural development was more stable and sustained when technologists and artisans from similar ecologies were enabled actively to cooperate with one another than when left to struggle in isolation. One of the earliest illustrations of the concept was demonstrated in a collaborative social forestry network in which foresters from close to a dozen low rainfall nations in the Middle East, North, East and West Africa participated.

Gilles Lessard, the Associate Director of Forestry, whose academic qualifications were in natural resource management, convened an informal meeting of the first participants in Dakar, Senegal early in 1972. Each forester gave a brief account of his nation's perceived priorities and opportunities, the resources needed and available to pursue them. In all nations, soil conditions had been debased by excessive removal of forest cover. Trees variously provided fuel, fodder for animals, fruits and nuts, materials of construction, commercially valuable commodities such as gum arabic, an exudate of *Acacia senegal*, and shea butter, used in foods and cosmetics after extraction and refinement from seeds of *Butryospermum parkii*. Several of the participants noted that trees provided soil stability and soil nutrients: deep-rooted trees pump nutrients from sub-soils that are concentrated in leaves which, when they fall, enrich surface soils. In some nations trees acted as sand-dune stabilisers and protected crops from strong winds when planted as peripheral barriers. Some years later, an Egyptian forester, Hosny el Lakany, employed *Casuarina* species as windbreaks to protect crops in a successful project, financed by IDRC, that reclaimed many hundreds of hectares of desert land for crop cultivation and livestock husbandry. He had learnt their use in desert reclamation from Coptic monks who had long experience in dryland agriculture.

To facilitate frequent exchange of information and materials, IDRC employed two forestry advisors, Abdul Seif el Din, a Sudanese fluent in English and Arabic, and Karim Oka, an Algerian fluent in English,

French, Arabic and Berber. The two foresters were given office space in Nairobi with funds to travel and convene meetings of the collaborators at least once each year. They provided a multilingual newsletter and when requested. IDRC provided monies to each of the collaborators to support priority projects. The social forestry programme eventually expanded to include some 20 projects supported by IDRC, most in arid and semi-arid ecologies of Africa and the Middle East.

In several of the francophone nations all forests and tree lots were owned by the government departments of Eaux et Forets. Trees and water owned by governments were frequently ill-protected and regarded as free to be exploited by all and sundry. Consequently, villagers harvested whatever trees and arboreal products they needed, while significant quantities were scavenged by goats and other migrant animals. It was suggested, and in two instances accepted, that all woodlots be declared the property of the nearest village community. As soon as the villagers owned the woodlots, they took pains to prevent depredation by animals and engaged in planting, protecting and conserving their woodlots efficiently and economically. In several locations, IDRC supported tree nurseries to generate species of particular utility to different ecologies and communities.

Other projects supported training of young rural people and establishment of small industries in timber processing including, in Tunisia, the production of paper from woody plants indigenous to the local ecology; in Mali the hybridisation of mangoes, and the establishment of small sawmills.

Agroforestry

During the 1970s, interest arose in integrating forestry with agriculture, alternatively named agrosylviculture or agroforestry. Soon after IDRC was created, there was pressure from some politically influential persons in the Canadian forest industry for the Centre to finance an international institute of forestry research in Canada. Eventually it was agreed that IDRC would explore the feasibility and utility of supporting an international programme to integrate forestry with crop and livestock production. Several international development agencies expressed interest and, in the later 1970s, after a somewhat difficult period, largely the result of inept politically influenced interference, a Centre for agroforestry research was established with headquarters in Nairobi and a small Board of Trustees drawn from several nations.

At first the organisation, named the International Council for Research in Agroforestry (ICRAF), followed the pattern of the Social Forestry network by providing advice mainly to African nations on various opportunities by which to cultivate and utilise arboreal species to protect food crops, to provide fodder for animals, to act as wind and insect breaks, and to provide soil nutrients in the form of leaf mulch. A search committee, led by a former Director General of the International Rice Research Institute, identified an exceptionally competent Director, Bjorn Lundgren of Sweden, and recommended that ICRAF begin its own research programme on land provided by the Kenya government. Given the qualifications of Gilles Lessard, it is interesting that the Director described ICRAF's programme as research and development to utilise land and related natural resources most efficiently.

During the 1980s other donor agencies provided support to ICRAF and the organisation was accepted as an associate member of the CGIAR. According to CGIAR records, ICRAF became a full member of the CGIAR in 1991. ICRAF is now a truly international research centre with programmes and projects in Africa, Asia and Latin America. Nonetheless, ICRAF metamorphosed from the IDRC social forestry programme and it was IDRC that gave it birth and nurtured it during its early, difficult years.

Local management of water

In the précis of his book *Water; Local-level Management,* David E. Brooks gives blunt advice: "There is one iron rule for managing groundwater and aquifer supplies: assume the worst". And then, from a man who has been in the forefront of this field heading an NGO

(Friends of the Earth) and worked as a consultant before joining IDRC, Brooks adds more positively: "Policy and research should shift their focus from enlarging supplies of water to managing demand". On the question of engaging local communities, he says: "Policy-making should always start by accepting social customs and cultural norms as given, but not sacrosanct". Finally a warning: "Devolution of the power to **manage** water (not just read meters and fix leaks) will not come easily. The forces to maintain a top-down approach to water are well entrenched and serve many power elites".

His way of thinking has informed a wide range of water projects supported by IDRC, that have taken place at all levels from international to community. Internationally, the Centre has worked for a dozen years in the Middle East and North Africa region to promote water demand management (WDM) through gatherings of ministers and officials to exchange knowledge and experience among their nine regional countries. Missions between countries have shown, for example, the Syrian water officials how Water Users Associations (local bodies) in Tunisia distribute drinking water and levy charges, and how Egypt deals with wastewater.

These gatherings have evolved into a Water Demand Initiative, a five year-programme (2004–2009) with the awkward acronym WaDImena, to reach out further to civil society with research, and to apply WDM strategies in particular rural areas, concentrating on women and the rural poor. WaDImena has a website (www.idrc.ca/WaDImena) with a trilingual glossary of WDM terms and a library that highlights lessons learned.

But respect for traditional knowledge is a major factor in tackling water demand problems. Farmers in the stony highlands of Yemen depended for years on intricate terracing to save water, and conserve fertile soils from erosion. Researchers studying why these terraces were now in disrepair learnt the obvious answer: maintaining them was brutally hard labour, and many men had left for the cities. Also, there was no clear agreement on cost-sharing between landlords and tenants, and no credit available to farmers to invest in water management. Ways were found to rebuild at reasonable cost and labour, and food production became profitable and attractive again. A comment from David Brooks: "Reviving traditional water management approaches can require both technical and policy ingenuity, but the rewards can be significant".

China provides the most striking example of how a determined (and sometimes desperate) community can take on management of its water resources, and reverse the top-down process of regulation. Centralised state control enforced by a cumbersome bureaucracy has long been the pattern in China, and one recent result has been that the economic boom in its eastern provinces have brought little benefit to parts of the interior. In particular, the mountainous Guizhou province in the southwest has remained one of the poorest; and Changshun county, set on sloping land that is mainly porous limestone is chronically short of ground and subsurface water. Shortage of water has meant low rice yields, little diversification of crops and degraded forests. It has added to the burden of women who walk miles to collect water for households.

A first attempt at water management was made in the early 1990s when facilities were rebuilt and maintained by the state. But there was little accountability of this poorly managed government project, and no local control. Politely put, the project had limited effectiveness. At that point, researchers at the Guizhou Academy of Agricultural Sciences (GAAS), funded by IDRC, decided to try a community-based approach, involving participatory decision-making about natural resource management. In 1995 its team began learning about and analysing traditional practices in a number of villages, and measuring the damage already done to the natural resource base there. (It was the same diagnostic approach described in the case-study on Farming Systems Research).

The GAAS team guided the academics and local residents through a collaborative process. There was debate on practical and technical matters – irrigation, reservoirs, pipes, new water sources – and strong attention to the social aspects of development. But the overall emphasis was on the *process* and on involving people in decisions about their own development. It was painstaking and lengthy, but the process worked. The people selected the technical solutions, and built the water systems; they took 'ownership' by working out regulations and agreed payments for maintenance and by electing (and paying) a manager to run daily operations.

Results reported have been remarkable. Farmers diversified crops and enjoyed increased yields. They grew fruit trees and berry bushes on the marginal land. Women probably gained the most. Relieved of hours of water gathering (it was now piped, a tap to every 50 households), they took charge of a peach orchard, and marketed the fruit. They overruled the researchers, who had suggested chestnut trees; and they impressed the deputy governor of Guizhou province, who had expected the scheme to fail, and who later admitted his officials had not thought of the kinds of regulations and management systems the villages established.

The idea that local initiatives can solve local problems is gaining ground in China. David Brooks should have the last words: "Decision-makers too often dismiss small groups and small solutions. They make a big mistake".

Essential health in Tanzania

There are some extraordinary, in fact appalling, statistics cited about Tanzania. Its national income in 2001 of $9 billion, divided among 35 million citizens, was roughly half what Americans spent that year on wallpaper. Its spending on health that year was US$11.37 a head, compared with $2,809 in Canada and $4,819 a head in the United States. A 'poverty head count' in 2003 showed 36 per cent of Tanzania's population living on less than US$1 a day. And there were an estimated 7,431 Tanzanians to every (ill equipped) health facility.

After independence in 1961 the government under President Julius Nyerere and his people made heroic efforts to build a health care system throughout the country. Under a unique social contract the rural people overcame their reluctance to move into modern villages, and the government provided pumped water, a clinic and a school. Graduates spread into rural areas, medical training centres were built. Health care was a public right.

By the mid-1980s the system was foundering. Its centrally planned management was unable to sustain the structure and to support efficiently the village dispensaries. And the people were coming under renewed assault from infectious diseases – malaria, tuberculosis and now AIDS. Payment obligations on international debt, coupled with sharply falling prices in export crops, cut deeply into funds for training health staff. Attempts to keep the system going by charging user fees for low-quality care only drove people angrily away from the whole system.

Even when the UN Global Fund and charities such as the Bill and Melinda Gates Foundation pledged substantial funds for health care in Africa, the health systems of Tanzania and other countries were weakened and had little ground-level capacity. The *World Development Report 1993: Investing in Health* [World Bank, 1993] argued that new investments should be based on evidence about the local "burden of disease" and be targeted proportionally. Increase health funding, certainly; but make sure it is allocated in a cost-effective way. The logic was good, but how to put it into practice?

Canada through the International Development Research Centre contacted several African countries to suggest a partnership to research this problem. Tanzania, intent on health care reform and planning to devolve the responsibility of budget management and health care delivery to local authorities, was the first to respond. The Tanzania Essential Health Interventions Project (TEHIP) has evolved as a true partnership between IDRC and the Ministry of Health, with Don de Savigny and Conrad Mbuya managing and coordinating research, and Graham Reid and Harun Kasale managing and coordinating the actual project.

TEHIP was launched in October 1993 with careful, broad-based consultations to design the project. There followed the stage of discovering how local authorities could get an accurate picture of the existing 'burden of disease' – which diseases imposed the greatest cost of lives (and years of life) lost in their area. It chose two rural districts, Morogoro and Rufiji, west and southwest of Dar es Salaam, with a total population of 741,000.

Most health data in Tanzania is admittedly inaccurate, being collected from clinics – while most people die in their homes. TEHIP therefore, sent researchers out on bicycles on a door-to-door survey, asking representative households whether anyone had died or been laid low with sickness recently, and with what symptoms. (Again the diagnostic approach: first, go and get the facts and views from the people affected). The findings were striking, showing some diseases were badly neglected. Malaria accounted for 30 per cent of the years of life lost in Morogoro, but had only 5 per cent of recent health budgets. Conversely, tuberculosis accounted for less than 4 per cent of years of life lost, but received 22 per cent of the budget.

IDRC funding offered a modest $2 a head top-up to the districts' budgets, but with re-allocation of existing funds only a part was at first needed in the transition to a more cost-effective system. Simple rules were followed: the cheapest treatments were offered first; drugs were ordered according to what was needed. (Under the old centralised system, all dispensaries received the same package of pills). Most important was the campaign to persuade people to use bed-nets impregnated with insecticide (these cost $3, locally made).

TEHIP remained in the background as the District Health Management Teams (DHMTs) planned their work and spent their money. The teams could work more efficiently unburdened with volumes of charts and paper, but armed with what Don de Savigny calls a "tool kit": a district profile of the 'burden of disease' with key indicators, graphs to cover trends in spending areas, health maps, community reactions and ideas, cost-tracking – all providing as simply as possible on computer diskettes.

The results were swift and dramatic. Adult mortality has dropped by nearly 20 per cent. Infant mortality fell by 28 per cent in one year in Rufiji, and the proportion of children dying before their fifth birthdays by 14 per cent, and in Morogoro annual under-5 mortality fell from 35 to 20 per 1000 within a few years. The teams achieved this by adopting another 'tool' that was devised by WHO: the Integrated Management of

Childhood Illnesses (IMCI) which meant (as de Savigny says) "addressing the whole child by identifying and treating a range of possible common illnesses rather than simply focusing on one disease at a time".

It can soon be more than a local achievement. DHMTs have been trained in half the country in the use of the burden of disease profile and the district health accounts tools. The United Nations has funded the 'roll-out' of the entire tool kit to 11 other Tanzanian districts, and CIDA has chipped in $7 million CAD to build on the results of TEHIP. Further afield, the Danish aid agency DANIDA is testing it in Ghana, and South Africa is showing a general interest in the TEHIP approach.

Making waves – and escaping the tsunami

During the 1980s India made such advances in information technology that it has become the world's second largest exporter of software from companies in Bangalore and Hyderabad. Yet the bulk of India's population lives in poverty, in rural villages. Professor M.S. Swaminathan, known in India as "the father of the green revolution" that ended recurrent famines, faced the nagging question: How can the information age bring benefits to those villagers who exist on less than US$1 a day?

A man of action, he set out to build a model for a pro-poor information revolution in rural areas, believing that a functioning computer in a village could be as valuable a tool for development as an irrigation pump or a community well. For there were in the early 1990s no models for how information and communication technologies (ICTs) could help the rural poor in India. The International Development Research Centre was the first donor, with project funding through the M.S.Swaminathan Research Foundation (MSSRF), to help him pursue this vision. It also provided MSSRF with information about how ICTs were used for development elsewhere.

There were plenty of technical challenges in the area, around Pondicherry in Tamil Nadu state, where the MSSRF researchers chose to work: an absence of any modern telephone infrastructure, and indeed a three-year wait for standard telephone lines. Also the team had a shoe-string budget. But they decided the plan was feasible, it was a simple matter of 'creative engineering' and an ordinary villager could learn how to use ICTs, given a chance. So the model should be community-owned, rather than placed in private hands.

The project director, Dr Venkataramen Balaji, a graduate of the Indian Institute of Technology at Kampur, devised a 'hub and spoke' model to connect several village knowledge centres (VKCs) to a central village by VHF radios, while the central base would be connected to the Indian telephone network and have Internet access. Each of these VKCs would have computer, printer and radio, and there would be solar power backup.

Dr Balaji also devised a computer programme that gives the 'biovillage' people instant access to the prices and demand in the markets they serve for the crops, livestock products and fish they produce. Also, computer graphics are used to instruct village people in new technologies, such as how to produce vermiform compost fertilisers and hybrid vegetable seed for sale to other growers.

A survey by the MSSRF team showed that people in the seven villages had a real thirst for information, provided it was locally personally relevant: men wanted daily weather reports for farmers and likely wave-heights for fisherfolk, available from a website of the US Naval Oceanographic Office. Women information on health issues and government programmes of family income supplements and so on. In return the Foundation required each VKC to stay open for several hours a day, guard the equipment well, guarantee access to members of the Dalit population (once known as 'untouchables') and ensure that half the trained volunteer operators were women.

Pondicherry was until soon India's independence a small French colony (as Goa was Portuguese), and the Territory's official languages are English, French and Tamil. For the sake of rural people, the personal computers feature Windows 95 and Microsoft software with Tamil fonts devised by the Indian government. For villagers who are illiterate, the centres download important information like weather reports as Real Audio files, then play them over speakers from outside the information shops. This procedure was crucial when the tsunami hit on 26 December 2004.

By chance on that day, Vijayakumar Gunasekaran, son of a fisherman from Nallavadu village who works in Singapore, was following the news of the earthquake off Aceh, Indonesia, as it built up the gigantic waves storming eastward. He decided to phone his home on the other side of the Bay of Bengal, and his sister told him seawater was already seeping into their house. He urged her to warn other villagers to evacuate their homes. She got two neighbours to break down the doors of the MSSRF centre (closed on that holiday), and they used the public address system to broadcast warning of the tsunami and sound the village siren. The whole village – 500 families and some 3,600 people – escaped inland, before the waves hit Vallavadu and destroyed 150 homes and 200 fishing boats.

The project has been making waves of its own. After seven years and with funding also from CIDA it now covers a dozen communities and some 50,000 villagers are making use of the wireless internet system. It has, as IDRC writer Keane J. Shore reports, prompted several examples of social change, particularly for women. Self-help groups based on the centres have begun supporting women with advice on medical issues, some of it provided by experienced elderly people. Some centres offer evening counselling sessions for women. Others have started microfinance groups and offer loans and training for women to start cottage industries. Women agricultural labourers, who are paid partly in grain use, keep informed about grain market prices from the centres. The list of initiatives that have blossomed is long. The centres themselves have built up large data bases.

In July 2004 the Swaminathan Foundation held a policy-maker workshop to share lessons from the project. Out of it came a national alliance with the ambitious aim of enabling up to 600,000 villages to be linked through rural knowledge centres by 2007. The Indian government has committed the equivalent of $28 million CAD to this 'Mission 2007'.

12

Political and Ideological Issues

Systems of governance

Isaac d'Israeli, father of the nineteenth century British Prime Minister, wrote: "Politics is the art of governing mankind by deceiving them". Mr. d'Israeli would consider his opinion amply vindicated by the misinformation propagated to justify the illegal invasion of Iraq. A distinguished Canadian who served for many years in both the House of Commons and the Senate expressed his private opinion that every politician has two ambitions: the first is to be elected, the second to be re-elected; all means employed to attain these ends are politically justified. Almost every critical observer referred to in the earlier text, from Barbara Ward onwards insisted that all national and international development – economic, social or technological – depends on political will and priorities.

For a government to declare its dedication to sustainable development is now fashionable, intended to demonstrate political and social enlightenment. Yet political policies are rarely sustained beyond the life of the party or authority that holds power. The recent election campaigns in Canada, the United States and India illustrate the diversity of political and ideological opinion among contestant politicians. A notable US senator, Daniel Patrick Moynihan sensibly stated that every politician has the right to express his or her opinions but not to present falsities as facts. In pursuit of their objectives to be elected or re-elected politicians make promises most likely to be attractive to a majority of voters. When elected into power the tendency is to promote policies beneficial to persons, political action groups and industries that have provided them with significant financial support. Since the poor nations of Africa, Asia and Latin America do not vote in North American elections, their urgent needs for assistance received scant mention during the recent election campaigns.

Unsustained political priorities

It is not unusual for a development agency to change its priorities when there is a change in its chief executive officer. CEOs derive immense satisfaction from putting their personal stamp on their agency's activities. Inconstant, unsustained support for development in poorer nations is illustrated by the fluctuation and overall decline in donors' financial commitments. In addition, programme priorities are subject to frequent change as new development fashions emerge. Over the past half-century, development slogans have included 'Freedom from Hunger', 'Freedom from Want', 'Women in Development', 'Integrated Rural Development', 'Sustainable Agriculture', 'Biodiversity conservation' 'Micronutrient deficiencies' and 'Responsible governance' the last now largely superseded by the US-driven 'War on Terror'.

International development priorities have frequently changed, some inspired by recommendations from many international conferences listed in the earlier text. Recommendations from the Rio Conference persuaded governments to focus attention on environmental issues, conservation of biodiversity, global climate change related to greenhouse gas emissions, and to arresting desertification. Though alleviation of poverty has been stressed as a constant and recurring need for many decades, earlier chapters illustrate that the gap between rich and poor continues to widen, while many millions struggle to survive in abject poverty. There seems little serious political will or imaginative sustained action to alleviate abject poverty wherever it exists. More recent development emphases include gender equality, human rights and responsible democratic governance.

A donor's vacillating policies

In 1987 the Canadian government of Brian Mulroney instructed an all-party committee of MPs to review and recommend on the nation's development assistance policies and programmes. The committee under William Winegard consulted with other development agencies, visited several recipient nations and compiled an imaginative, thoughtful commentary with recommendations. Most notable was a recommendation that the national aid agency's facilities be decentralised, that more aid administrators be stationed in recipient countries and developing regions. The Winegard committee sensibly stated that administrators who live among developing communities are more aware of current and changing needs and opportunities, better able to define objectives and to monitor project progress, than are administrators who make only brief, transitory visits from a far distant headquarters (Winegard 1987).

The government and CIDA its aid agency acted on the decentralisation recommendation and transferred many of its staff to overseas locations. Subsequent evidence showed notable improvements in project formulation and monitoring.

In 1990, the same Progressive Conservative government hired a private consultant agency, Montreal-based Secor, to undertake another review of its development assistance policies, programmes and practices. The consultants concluded, in its report, *Canadian International Development Agency: Strategic Management Review,* that decentralisation had raised the cost of delivering aid, and that the agency should reduce its "labour-intensive activities". In consequence, decentralisation was put into reverse, many overseas aid administrators being transferred back to the national headquarters.

The hired consultants' stated objective was to reduce the cost of delivering aid. If the objective of aid is simply to transfer money from the donor to the recipient with little concern for the effectiveness and benefits to the poor that ensue, then the cost of delivering aid can indeed be drastically reduced. But as Pearson and others have so often stated, for aid to result in productive sustained development requires long patient, persistent, sensitive, competent study of the differing conditions in each community to be assisted. There are no quick and easy development processes.

Misguided by their evident inadequate understanding, the hired consultants failed to recognise that alleviation of poverty by enabling poor people to gain sustainable employment is essentially labour-intensive.

Opportunities for employment and resources available differ significantly among the world's poverty-stricken communities. As has been demonstrated in India, poor rural women with little formal education can be trained, equipped and motivated successfully to own, manage and operate small rural agribusinesses. To do so necessitates a thorough systematic assessment of prevailing conditions, opportunities and resources in each community that is to be assisted (*see* Chapter 11).

Comprehensive assessments and subsequent support and stimulation of rural employment are highly labour-intensive. It is not improbable that bilateral aid agencies' failure to alleviate poverty by support of projects to create employment opportunities is attributable to their preference for large capital-intensive, rather than small labour-intensive, projects. Less effort is required to write a single cheque for $10 million than one hundred cheques each for $100,000. Furthermore, such small rural employment projects make little demand on tied-aid resources and advisory services. So long as bilateral aid agencies opt for large capital-intensive projects with heavy tied-aid provisions, there is little hope that chronic poverty will be appreciably alleviated.

Concepts of democracy

What constitutes democratic governance differs markedly among political régimes motivated by divergent ideologies. Concepts of 'democracy' have differed and changed since first proposed by the ancient Greeks. The word 'democracy' is derived from a Greek word δεμοκρατια which may be interpreted as 'rule by the people'. Modern dictionaries define democracy as: government wherein sovereign power is exercised directly by the people, or by officers fairly elected by all the people they are to govern. Modern perceptions of democracy assume that equal justice will be available to all citizens. Some self-proclaimed democracies are more accurately described as autocracies Greek: αυτοκρατια (autokratia) where people are ruled by self-sustained power and self-constituted authority, typical examples being military dictatorships and régimes that forcibly preclude all legitimate political opposition. Vote rigging and other devious anti-democratic actions to gain and retain power by illegitimate means are by no means confined to isolated instances.

Some that claim to be democracies are more correctly designated 'plutocracies' Greek: πλυτοκρατια (plutokratia) which means 'rule by the power of wealth'. Nations wherein election to presidential office, to national parliaments or state governorships is effectively restricted to an affluent minority who can afford or acquire from wealthy supporters the money to cover the prohibitive cost incurred in pursuing elective office are *de facto* plutocracies (πλυτοκρατια), not democracies. It is estimated that, in total, several billion dollars were disbursed by the candidates who compete for the US presidency and vice-presidency, and by their supporters. Immense financial resources are demanded in order to compete for a seat in the US Senate or House of Representatives. Such extravagant costs of campaigning automatically exclude the overwhelming majority of citizens from seeking election to national or often State and local public office.

Perceptions of ideal government varied among the ancient Greeks. In Athenian democracy, administered by wealthy, influential aristocrats, slaves who had no right to vote or participate in government were accepted as legitimate members of the community. Though ostensibly slavery was abolished by Britain in 1833, and in the United States 30 years later, *de facto* slavery is still evident where governments turn a blind eye to illegal immigration by desperately poor people, who are then employed as cheap labour in domestic service, agriculture and textile sweat-shops.

Plato's concepts of democratic government and justice had more to do with property rights than with equity for all inhabitants. Plato sensibly proposed that all governments should be led by philosopher kings: scholars who could study, think and plan rationally. Presidents, prime ministers and leaders of governments possessed by rational scholarly intellects are not widely evident among present day powerful and affluent nation. Aristotle, whose perfect man was the golden mean between extreme behavioural characteristics, did not believe that all citizens should be granted equal political rights. In his *Politics* Aristotle declared that good government must ensure what is good for all citizens, that good and bad government are defined by the ethical dispositions of those in power, not by any form of written constitution. Aristotle distinguished between his definitions of oligarchy and democracy: the former exists where the rich govern without concern for the poor; democracy puts power into the hands of the poor who disregard the interests of the rich.

An unresolved controversy among different perceptions of democracy relates to collective and individual wellbeing – the greatest good for the greatest number, versus the unconstrained right of the individual to attain power and wealth. Jeremy Bentham, the eighteenth century leader of the 'Philosophical Radicals' conceived democracy as a system that provides the greatest happiness for the greatest number, in contradistinction to the seventeenth century doctrine of the Rights of Man which the economic historian R.H.Tawney in his book *Religion and the Rise of Capitalism* describes as everyone's having the natural right to do whatever makes for their greatest personal advantage. The weakness in Bentham's philosophy is that people and communities derive pleasure from widely divergent activities and sources, some active in composing and performing music, some passive in enjoying music performed by others. The Rights of Man places the rights of the individual over the rights and needs of the bulk of society and is the guiding principle in nations and societies where the unrestricted acquisition of wealth and power is the *sine qua non* of exemplary success. Nonetheless, basic human rights of free speech, open assembly, free expression of religious and political belief, and equitable treatment under a just legal system are vital to every civilised nation. The diverse philosophical concepts of human purpose, governance and democracy are elegantly described and discussed by Bertrand Russell [1967].

In a book written during World War II, H.G.Wells asks: "What is democracy?" [Wells 1942]. He writes that democracy is associated with freedom, liberty, rights of individuals and of communities to self-government, "the subordination of the State to the welfare of the individual". He goes on to suggest: "There is no such thing as absolute freedom or absolute servitude". Democracy is dynamic and subject to frequent systematic review and renewal. Unrestrained economic growth with advantages restricted to a privileged few imposes a grave infringement on human liberty. Democracy means: a. economic justice and sufficiency for all people, every individual's basic needs must be satisfied; b. freedom of speech, association and individual expression of opinion. Wells contends that only through a world democracy with a world democratic government with powers of legal authority is there hope for the ultimate survival of our species. Domination of weaker nations and communities by the most affluent and powerful will result in widespread violent reactions by those who are dominated and oppressed. Violence begins when rational mutually sympathetic debate becomes impossible, a proposition amply borne out by people of Middle Eastern origin stimulated by the invasion of Iraq and the biased and bigoted political decisions inflicted on Iran and the Palestinians.

A world organisation with authority to adjudicate conflicts between and among nations, to enforce equitable justice for all humankind was the vision of those who created the United Nations Organisation. The UN Security Council is charged with the primary responsibility of maintaining international peace and security. Sadly, the Security Council is inhibited in fulfilling this responsibility by the power of veto exercised by the permanent members, a power exercised whenever a proposal conflicts with a self-centred national interest. The International Court of Justice is effectively emasculated by the refusal of certain powerful governments to recognise the Court's jurisdictional authority.

Some powerful nations that claim to promote democracy are given to repudiating results of democratic elections when those elected proclaim social and economic philosophies that differ from those of the more powerful. The democratic elections of Allende in Chile and Hamas in Palestine illustrate the point.

In his final speech as Secretary General of the UN, Kofi Anan said, "Poor and weak states are easily held to account because they need foreign assistance. Large and powerful states where action have greater effects on others, can be constrained only by their own people".

What is the purpose of human existence?

Generations of philosophers have meditated on the origins, nature and purpose of life on Earth; were humans created to fulfill some divinely ordained purpose or is human existence the result of a series of random genetic evolutions? In his book *Das Kapital* Karl Marx insisted that philosophical studies are not simply to speculate on the world and patterns of human existence and behaviour, but to alter them. However, Marx was possessed of an indolent and unsystematic intellect, incapable of proposing in detail how his economic and social theories could be translated into benign productive systems of governance.

Alternative philosophies have been classified as: a. hedonistic – the primary purpose of life is the pursuit of pleasure; b. intellectual – the acquisition and elucidation of knowledge should be the primary purpose; c. pragmatic/egotistical – life should be devoted to an insatiable pursuit of wealth and power. Apart from such remarkable exceptions as St. Francis of Assisi who regarded a life of abject poverty, suffering and avoidance of all pleasure as a guarantee of eventual eternal paradise, most people would aspire to some measure of all three. Money and material possessions are essential for food, clothing, shelter and modest access to recreation and entertainment. Knowledge and the skills engendered by education and training are essential to productive employment. Knowledge of arts, music and literature are sources of pleasure and diversion. Employment requires skilled knowledge.

While few would opt to live in a community of clones as described in Aldous Huxley's *Brave New World* or George Orwell's *1984,* a national or international state in which the rich are uncontrollably free to acquire greater riches, while the poor descend into more dismal poverty is inevitably unsustainable. While some instances of modern terrorism are attributable to extreme religious or political bigots who claim divine or despotic authority to impose their dogmatic beliefs on everyone else, many more result from desperation of unrelenting poverty, inequity and oppression. Wells [1942], echoing Socrates, contends that every energetic young person is a potential troublemaker if he or she is denied gainful employment and a sustained means of livelihood.

Social welfare in a secular society

Most major religions proclaim the responsibility of the powerful and wealthy to take care of the poor, the infirm and the disadvantaged. In our more secular societies people expect their governments to provide social programmes to alleviate poverty, provide health care for all, support and sustain all citizens throughout their old age. Among the more affluent nations of North America, Europe and Oceania, while birthrates are falling people are living longer, leading to an inexorably rising proportion of old, retired people, with a declining proportion of younger men and women able to work, earn wages and pay taxes. Older people being more susceptible to degenerative diseases have need of more health care than younger folk. In light of the rapidly rising costs of diagnostic devices, of developing, evaluating and marketing new therapeutic drugs, of providing curative and palliative treatment to a growing ageing population, free health care for all will soon become economically unsustainable even for the wealthiest nations.

Alternatives, all highly contentious, include: (a) raising of conventional retirement ages to permit people to work over more years;

(b) totally free health care to be provided for only the poor, with people who can afford to do so being required to contribute to the health service they enjoy; (c) higher compulsory contributions from all wage earners to superannuation, retirement, pension and health care funds; (d) progressively increased taxation. Few politicians who hope to be elected or re-elected display the intestinal fortitude to propose unpopular policies; they prefer to leave inevitable financial and social difficulties for their successors. Politicians *pace* Plato are rarely scholars given to long-sighted systematic planning; more often their imagination stretches little further than the next election date.

Religion and politics

In a disturbing article in the journal 'Foreign Affairs', W.R. Mead discusses the extreme influence exercised upon the US government by Christian fundamentalists and evangelists. He writes: "... Fundamentalists are downright hostile to a world order based on secular morality and global institutions such as the United Nations ... (Their) vision is not hospitable to gradual progress towards a secular utopia driven by technological advances and cooperation (among) intelligent people of all religious traditions" (Mead 2006). The name 'religion' embodies the Latin verb 'ligare': to bind together so that members of a particular faith or religious sect are bound together by a common religious dogma. There is a melancholy historical litany of inter-religious conflict, both between different religions and between sects of the same supposed religion. Much of the present turmoil in the Middle East results from the claim of certain believers to be God's chosen people with unique rights and ownership of particular territories.

The UN Charter and all rational definitions of human rights proclaim the right of all people to adopt the religion of their convictions. But they do not permit those of one religion to impose their doctrines on others by force of arms. Militant declarations by members of one religious sect provoke militancy among those of other faiths who perceive themselves as threatened. There can be little hope of peace so long as powerful politicians are influenced by aggressive believers in any particular religion or sect. There can be little hope of sustainable development for all humanity so long as inter-religious and inter-sectarian violence prevent others from worshipping according to their beliefs. Fundamentalist evangelism by any religious sect can lead only to conflict and bloodshed.

Unsustained and unsustainable political systems

Between 1776 and 1788 the seven volumes of Gibbon's *The Decline and Fall of the Roman Empire* were published [Gibbon 1985]. Gibbon's scholarly work describes how a nation that for centuries ruled the known world, that enacted a justice system subsequently adapted by many Western nations, gradually disintegrated into relative insignificance. Gibbon states somewhat pessimistically: "History is little more than a register of the crimes, follies and misfortunes of mankind". The European fascist hegemonies were destroyed by World War II. The Soviet empire fell apart largely because the Marxist dream of rule by the proletariat degenerated into vicious oppression of a defenceless majority by a powerful political minority.

During the second half of the twentieth century the colonial empires dominated by European nations gradually dispersed into self-governing states. Sadly, many were too small and possessed too few resources to survive as economically sustainable entities. Several attempts at regional cooperative associations, such as the Caribbean federation and the East African Community in which neighbouring nations were intended to pool their resources collapsed because of chauvinistic politicians unwilling to accept dilution of personal power in the interests of cooperative viability.

At the time of writing it is the abundant wealth, military power and assumption of divine right by one isolationist nation that dominates the rest of the world, a nation whose inhabitants and industries consume a disproportionate amount of the world's limited resources, and assume that all disputes are most conveniently settled through the barrel of a gun. History teaches that no nation, however powerful, remains dominant for ever. It is by no means inconceivable that within 50 years the Peoples Republic of China, hardly redolent of democracy by any definition, and/or India, the world's largest true democracy, will overtake the United States as the world's dominant economic and military power.

China's economic growth and industrial capacity are expanding at an astonishing rate. Its population is almost four times the size of the United States, it is rapidly acquiring the most advanced of industrial and military technologies. Its dictatorial system of governance that controls all sources of information ensures absolute secrecy and denies the outside world any comprehensive knowledge of its overall development. There seems little likelihood that the existing régime will be superseded by one more democratic, more open to external scrutiny. Because of the wisdom of its first rulers after independence, India has developed an enviable record for scholarship, for research and innovation in many areas of modern science and industry. Sadly in both India and China the gulf is widening between people with increasing wealth and the millions who suffer squalid, abject poverty.

Some observers express the hope that the European Union will become an exemplary benign force for international harmony and cooperation. At present the EU is little more than a loose trading federation of fractious members, some more dedicated to retaining their political, economic and social sovereignty than being less chauvinistic participants in an international cooperative enterprise. The distortions to international trade brought about by the EU Common Agricultural Policies, dictated by a few nations' farming communities, illustrate the present lack of acceptance of cohesive common policies. It is unfortunate that the EU has been pressured into major investments in armaments. One had hoped the EU would have devoted its resources to more humane purposes, particularly to ensuring greater equity and justice in the Middle East.

Nations, which despite insignificant military muscle at one time exercised a benign influence in international affairs, may lose that influence as their political philosophies and priorities become unimaginative and myopic. Cohen [2003] discusses what may be described as the decline and fall of Canada's international influence since "the Golden Age of Canada's foreign policy". During the 1950s, when Winston Churchill designated Canada "the lynchpin of the English-speaking world", brilliant scholarly leaders such as Lester B Pearson gave Canada an enviable international standing as peace-maker, peace-keeper and generous donor, a nation noted for cooperative, constructive diplomacy [Cohen 2003]. Granted, the world is now more complex and heterogeneous than in 1950 when the UN had fewer than 75 members, since then grown to 191; when over the same period the Commonwealth members grew from 8 to 54. The now expanding European Union with a present membership of 27 nations did not exist in the early 1950s. Nevertheless, relatively small, non-militaristic nations such as Canada and the Scandinavians, and nations like India, rapidly developing in industrial technologies and economic growth, can influence international behaviour if they elect leaders of creative vision with the will and determination to promote compassionate international cooperation rather than conflict, leaders who recognise that peace and security demand respect for philosophies and values different from their own cultures and ideologies.

The world has urgent need of leaders of Pearsonian persuasion, who accept the rights and responsibilities set out in the Atlantic Charter; who recognise that ideological differences are more sensibly settled by sympathetic discussion than by force of arms; who enact legislation to reduce excessive consumption of energy and water, legislation that restricts pollution of the global atmosphere, destruction and degradation of scarce arable land. As Sir Martin Rees, Britain's Astronomer Royal, warns in his publication *Our Final Hour*, in the absence of a substantial reduction in resource consumption and environmental degradation, "the odds are no better than 50:50 that our present civilisation on Earth will survive to the end of the present century".

India's potential in world affairs

In this author's opinion India offers the best hope of exemplary governance, as well as social and economic development. India, already the world's largest parliamentary democracy, will soon be home to the largest national population. As a result of its enlightened investment in scientific and technological development from the time of independence, India's progress in agriculture, industrial biotechnologies and information industries has been exceptional. India's wheat production has risen from 6 million tonnes in the 1950s to 70 Mt in 2003. The retail value of processed foods today is more than 1000 times the value in 1962. Prophylactic and therapeutic drugs produced by a vibrant indigenous industry are notably cheaper than their counterparts in more affluent nations. It is estimated that more than 25 per cent of the world's leading computer programmers were born in India. Recent encouraging statements indicate that India is prepared to convert from being a recipient of aid to become a contributor, and will assist poorer nations to improve their standards of living by sharing India's immense scientific, technological and developmental experience.

While India's industrial technologies and prosperity among a growing middle and upper-income class demonstrate admirable progress, a sizable proportion of Indians still exist in abject poverty. As successful Indian industrialists amass enormous wealth, millions of poor Indians are denied access to the basic necessities for a healthy life. India can become an exemplary model for national governments throughout the world if and when it combines and integrates its scientific and technological ingenuity with the ethical and social principles so eloquently propounded and demonstrated by Mahatma Gandhi.

Among his many wise observations Gandhi wrote: "Economics that hurt the moral wellbeing of an individual or a nation are immoral and sinful. Economic pursuits that commit one country to prey upon another are grossly immoral". "To a people famishing and idle, the only acceptable form in which God dare appear is as work and a promise of food as wages". Together with Rabindranath Tagore, the Nobel laureate for literature in 1913, Gandhi hoped and prayed for "a magnificent harmony among all human races". Few humans who lived so simply and humbly as Gandhi were accorded the glowing acclaims and tributes that flowed from national leaders and persons of influence around the world following his brutal and senseless assassination. To quote only a random few: "He was the friend of the poorest, the loneliest and the lost"; "No other living human being so forcefully demonstrated the power of the spirit over material things"; "No one had ever confronted adversaries at home and abroad and won so many victories with the simple weapons of kindness, honesty, humility and non-violence". Perhaps the most prophetic, from a senior British politician: "Gandhi's greatest achievements are yet to come." [Fischer 1954; Prabhu 1960].

In his eulogy to Julius Caesar, Shakespeare's Marcus Antonius said: "The evil men do lives after them; the good is oft interred with their bones". Though there are Indians and others around the world who seek to emulate and advocate Gandhi's principles of kindness, honesty, humility and non-violence, they exert a declining influence in a world dominated by avarice and the pursuit of power. If India were to reaffirm, practise and promulgate the good exemplified in the Mahatma's life, Gandhi's greatest achievements could yet be realised. One of Gandhi's most profound aphorisms – "Our planet's resources are sufficient to satisfy everyone's need but not everyone's greed" – seems to inspire little beneficial influence over many leaders of governments and industries. When India's system of governance and national prosperity is sustained by its exceptional industrial and technological progress, dedicated to and philosophically inspired by Gandhian principles it will truly become the world's greatest democracy.

Political participation in national development

What should be the role and responsibilities of governments in national economic and industrial development is highly contentious. Politicians who overtly oppose 'Big Government', and insist that the private sector pursue its purposes free of government interference, enthusiastically support massive diversions of tax revenues to acquire armaments, and to bail out failing airline companies and other unprofitable enterprises. The vast and complex subject of governmental involvement in economic, industrial and social affairs is thoughtfully reviewed in the third of a trilogy of related publications [Galbraith 1973].

Government support for commercial and industrial development varies markedly among nations in both magnitude and modalities. Among highly industrialised nations, governments offer support for research and development in a variety of patterns: direct grants, advantageous taxation, facilities for industrial research within government or quasi-government laboratories. Soon after independence the government of India established the Indian Council of Agricultural Research and the Council of Scientific and Industrial Research, which now administers more than 40 research institutes, each devoted to a particular industry. The Nehru government's enlightened encouragement of science, agricultural and industrial technological development has resulted in admirable economic and social benefits to the nation, and progressive stimulation of indigenous industries. Today, an impressive number of Indian industries maintain sizable, productive research departments.

Regrettably, the Indian example cannot be emulated by smaller nations whose resources are more limited. In several poor African countries, UN and bilateral agencies created and equipped industrial research institutes which, when the aid came to an end, could not be sustained. Equipment that required spare parts from abroad, to be purchased with foreign exchange, could not be repaired. Such industries as existed could not take advantage of technological innovation, with the result that many of these institutes are now barely functional. As is illustrated in the case studies entered under the title "Employment in rural agribusiness" in the previous chapter, small agribusiness industries are better stimulated and sustained by accessible technical and financial advisory services than by innovative research.

Governments and industrial research and development

Accurate estimates of government and industrial research investments are constrained by divergent perceptions of what constitutes 'research'. Some years ago this author proposed as a working definition: "Systematic progress from the known into the unknown". A manual published by the OECD is "to assist member countries who collect and issue national R & D data" [Frascati 1981]. The *Frascati Manual* proposes the following definitions:

"*Basic Research* is experimental or theoretical work undertaken primarily to acquire new knowledge ... without any particular application in view".

"*Applied Research* is original investigation undertaken to acquire new knowledge directed primarily towards a specific practical aim or objective".

"*Experimental Development* is systematic work, drawing on existing knowledge from research or other sources, directed to producing new materials, products or devices ... or to improving those already produced".

The *Frascati Manual* describes and discusses other technical and investigative activities related to research and experimental development. Though the Frascati definitions allow considerable latitude for interpretation, the *Manual* offers a useful guide to assessing the cost and evaluating the output of government financed research and related activities. Much of what is designated industrial research is in fact product and process development: the provision of goods and services that can be profitably marketed to identified customers. Few companies engage in basic research, as defined by Frascati, while only companies whose products demand high prices can afford to pursue long-term applied research.

Irrespective of other priorities, government support for agriculture, food and drug development and monitoring is a critical responsibility. Almost every nation is in some degree dependent on a national agricultural system; in most jurisdictions the government is the main contributor to agriculture, particularly where trade in agricultural commodities is important to the national economy. The annual international trade in agricultural commodities is valued at *ca* US$600 billion; annual subsidies to agricultural exports by the EU, United States and other affluent nations exceed $350 bn.

It is every government's responsibility to ensure food and health security, to protect its citizens from deception and fraud, from insult and injury to their health. Governments must therefore enact and enforce fair trading legislation, and create a capacity to ensure that all foods and drugs sold and dispensed meet specified standards, a capacity that entails maintenance of well equipped laboratories and employment of professionally qualified scientists, requirements beyond the resources of many small, poor nations. Regional food and drug laboratories, to serve several small nations are highly desirable but little in evidence.

Future demands on government interventions in development

As stated earlier in the text, the rapidly diversifying needs and demand, the fast rising costs of ensuring food and health security for expanding urban communities necessitate ever increasing government participation in economic, social and industrial development. Among the industrialised nations, it is the ageing populations who need more expensive health care that present increasing difficulties. Among poorer nations, in addition to the spread of debilitating diseases, the need for employment by predominantly young people is of extreme urgency. For neither group of nations is there a universal remedy; each nation, each community is distinct and different. Their needs and opportunities must be addressed individually. They will be neither solved nor satisfied by convening more massive international conferences that simply reiterate the same old broad generalisations.

John Bunyan wrote 'Every vat must stand on its own bottom.' It is the responsibility of the rich and powerful to enable poorer nations democratically to decide and determine then own political destinies.

13

Ethics, Communications and Education

Ethical issues and concepts

Bertrand Russell describes ethical philosophies as a vast area of speculative uncertainty intermediate between science and theology [Russell 1957]. MacIntyre writes that, whereas physics and mathematics are based on scientifically verifiable facts, the only information available to ethical philosophers are opinions [MacIntyre 1966]. In his *Discourse on Method* René Descartes gave as his opinion: "There is nothing in the sphere of philosophy and ethics that is not in dispute; nothing that is above doubt" [Descartes 1966].

Over many millennia rulers, philosophers and scholars have proposed their concepts of what is ethical and moral. Though some contemporary writers distinguish between 'ethics' and 'morals', the two are essentially synonymous, the former being derived from 'ηθικοζ' (ethikos) which a Greek dictionary defines as 'ethical, moral', the latter from '*moralis*' which a Latin dictionary defines as 'moral, ethical'. The *Oxford Dictionary* defines 'ethics' as 'the science of morals, of human conduct and duty'. It defines 'morals' as 'pertaining to human behaviour and conduct, to distinguishing between right and wrong, between goodness and evil'. Aristotle regarded ethical behaviour as that which is conducive to the greatest happiness. Others have defined ethical standards in terms of justice, truth, equity, right and wrong, protection of the weak from oppression by the powerful. The Stoics contended that to be ethical is a divinely ordained duty, a virtue that is its own reward. Hammurabi's *Code* in effect defines ethical standards, its stated purpose being: "To cause justice to prevail, to destroy the wicked, to prevent the strong from oppressing the weak, to protect the health and welfare of the [Babylonian] people". Over many centuries philosophers have debated whether actions and activities be judged ethical or unethical by a motivating sense of duty or obligation (deontology) or by resultant consequences (teleology).

Though the distinguished Professor of English, C. S. Lewis believed that a universal code of ethics exists, in the present world such is not so. There are nations that permit infanticide; there have been instances of government-sponsored genocide; some nations torture prisoners and impose the death penalty for certain crimes. Some governments consider induced abortion to be a mortal sin, while others contend it is a woman's right to decide to terminate an unwanted pregnancy. Despite disparities among religious, philosophical and philological interpretations, from the time of the ancient Egyptians and Babylonians certain ethical/moral principles have been accepted by socially sensitive societies and scholars: people should not commit murder, steal or be dishonest; the rich should extend compassionate care to the poor and sick; the powerful should not oppress the weak. Sadly, while dominant governments and their affluent citizens may pay lip-service to these ethical/moral principles, their unrelenting pursuit of greater wealth and power would persuade Francis Bacon to amend his aphorism from "*Nam et ipsa scientia potestas est*" to "*Nam et ipsae divitiae potestas est*". Wealth not knowledge is now pursued as the means to national and individual power.

For comprehensive scholarly reviews, readers are referred to Russell [1957, 1967]; Nowell-Smith [1959]; MacIntyre [1966]. This chapter will be confined to certain contemporary issues deemed ethical and related to biotechnologies and biotechnological industries, subjects addressed at greater length elsewhere [Hulse 2002].

Ethics in biotechnologies

Five issues that relate to biotechnological ethics will be discussed; Bioethics in medicine; Genetic modification of organisms; Human embryonic stem cells; Functional Genomics; Bioindustrial ethics.

Bioethics in Medicine

More than 2,000 years ago the Greek Hippocratic School of Medicine defined the Hippocratic Oath, intended as a code of ethical practice for all physicians and surgeons. The Oath specifically forbids euthanasia and requires medical practitioners to protect their patients from harm and injustice.

During the eighteenth century, many medical colleges were founded in the United States by Scottish Presbyterians who appeared to believe that strict adherence to Calvinistic religious doctrine was sufficient to ensure ethical medical practice. Unconvinced that Puritanical piety was the sole essential guiding principle, in 1847 more than 120 physicians met in Philadelphia and founded the American Medical Association (AMA) specifically to define and adopt a code of biomedical ethics. It included protocols for relations between physicians and patients. More recently the AMA has laid out bioethical responsibilities in health and medical services to include: (a) to save and prolong human life; (b) to restore health to those who are sick; (c) to relieve pain, physical and psychological symptoms of disease and distress; (d) to educate and counsel the public at large on how to maintain and enjoy good health.

Though American lawyers amass substantial fortunes from prosecuting what they proclaim as medical malpractices, it is this author's belief that most physicians and surgeons conscientiously adhere to the AMA ethical principles. Mischievous malpractice suits by acquisitive lawyers significantly raise the cost of health care because of high premiums charged by insurance companies to protect medical practitioners against costly litigation. There would seem to be a strong case for an enforced code of ethics by which to control the legal professions.

Genetically modified organisms

Organisms of the same botanical or zoological species reproduce and propagate their species by sexual reproduction. In the case of flowering plants a male gamete contained in a grain of pollen unites with a female gamete in the recipient plant's ovule to generate a fertile embryo. Mendelian plant breeding begins by crossing two sexually compatible genotypes followed by many years of selection to produce homogeneous, genetically stable progenies that combine and display desirable properties inherited from the parents. This long time-consuming process can be short-circuited by direct transfer of genes that encode for specific desirable properties. While Mendelian methods permit properties to be combined into sexually compatible species, modern transgenic techniques transfer genetically controlled properties between unrelated sexually incompatible species and organisms. For example, resistance to pests and parasites that have evolved in wild plant species can be genetically transferred and stabilised in unrelated cultivated crop species, thus reducing the need for crop protection by toxic chemical pesticides. Similarly, tolerance to salinity demonstrable in mangrove plants can be transgenically transferred into cereal crops naturally intolerant to salinity, permitting the genetically modified crop to be cultivated on saline soils.

The first potentially beneficial GM experiments were designed to transfer pest resistance genes from wild species into cultivated rice (*see* Chapter 7). Unfortunately an international seed company adapted transgenic techniques to induce into maize and soybeans resistance to the company's proprietary herbicide, thus encouraging North American farmers to increase weed-killer applications without damage to the cash crops. This reversed the original intention to reduce the incidence of toxic pesticides by increasing the use of chemical herbicides. The subsequent discovery that pollen and seed from herbicide-resistant genotypes could be carried by birds, insects and air currents to contaminate non-GM crops of like species led to an immense outcry against GM crops, further aggravated by uncertainties about the possible effects of the biochemically changed GM crops on human health.

Widespread opposition to all genetic modifications was stirred up by mischievous journalists who described them as "Frankenstein foods". Much of the opposition was scientifically irrational, ignoring the facts that: a. genetic modifications of organisms have occurred naturally over many millennia as genes were translocated between and among organisms by bacteria, viruses and other vectors; b. genetic modifications may be used to bring about widely different biological objectives. The unhappy state of affairs was made worse by people who tried to disguise genetic modification by naming it 'biotechnology' (singular), which led to adversaries condemning biotechnologies in general. As described in the earlier text, some EU agencies seem to regard 'genetic modification' and 'biotechnology' as exclusively synonymous.

Mistrust of GM crops is exacerbated by a prevalent suspicion of novel, unfamiliar foods and food processes. Food preservation by irradiation was first sponsored by the US Congress and Department of Defense. In 1953 the Director of the US irradiation programme wrote: "For the sterilisation of foods, methods may be evolved that utilise waste products from the manufacture of atomic bombs". Despite the World Health Organisation's assurance that, at prescribed doses, irradiation poses no greater hazard to human health than many more ancient preservation processes, food irradiation has never recovered to become wholly acceptable in the eyes of consumers. Other novel processes, including genetic modifications, are similarly suspect to many consumers.

Indisputably unethical is potential misuse of GM in what is labelled Genetic Use Restriction Technology (GURT), described in demotic terminology as Terminator Gene Technology. GURT embraces methods designed to restrict use of genetically modified plants, the second generation seeds of which are sterile. The sterile seeds prevent farmers from saving a portion of harvested seed for future planting, thus condemning farmers each year to purchase fresh seed from the seed company. One modification of GURT so modifies the crop that any genetically induced improved property does not function until the plant is treated with a chemical substance, also sold by the seed company.

Members of the UN Convention on Biological Diversity have condemned GURT and the restrictions GURT would impose on farmers who traditionally retain some harvested seed for future planting.

Publicists bent on arousing alarm among consumers are doubtless aware that, scientifically, it is easier to identify potential causes of acute or chronic toxicity than to demonstrate that a food will never under any circumstance impose insult, injury, distress or discomfort to any living person. Different individuals suffer intolerance or are allergic to particular food ingredients that cause neither disease nor discomfort to most others. Lucretius had valid reason when he wrote: "*Quod ali cibus est aliis fuat acre venenum*": What is food for one person may be bitter poison to others.

Most innovation entails some element of risk. It is incumbent on all scientists who devise and develop novel foods or food processes to ensure that the resultant novelties pose no scientifically determinable hazard to the health of consumers. Plant breeders should seek the active assistance of biochemical analysts and toxicologists to ensure that transgenically induced biochemical changes are unlikely to cause insult or injury to consumers' health.

Potential benefits from genetic modifications

Scientifically controlled genetic modifications of diverse organisms offer exceptional future benefits to agriculture, food and drug developments, to the treatment of obnoxious waste, contaminated soil and water. GM can enhance nutritional and functional properties and extend the storage life of fruits and vegetables. Many useful enzymes can be synthesised and isolated from GM microorganisms. Enzyme catalysed reactions are generally more conservative of energy and protective of environments than chemical catalysts. GM organisms are effective in removing or neutralising toxic and offensive contaminants from soils, ground and surface waters, and are able to convert biological wastes into animal feed, organic fertiliser and renewable energy.

Pharmaceuticals synthesised by cultured GM organisms include vaccines, hormones, immune regulators and therapeutics that control haematological and vascular disorders. Vaccines from GM viruses include whole virions (for poliomyelitis), split vaccines (for influenza), and isolated antigens (for Hepatitis B). Canadian scientists were first to extract an insulin precursor from a GM bacterium. Factor VIII to control blood clotting can be synthesised in the milk of GM cattle. Anti-clotting and anti-haemophilic factors are industrially produced from cell culture of GM organisms. The diversity of medically and nutritionally useful biochemicals that may be synthesised by cells cultured from GM viruses, micro-organisms, higher plants, insects and mammalian tissue is beyond imagination. Bacteria and viruses can be cultured for metabolite synthesis or as vectors to transport genes between organisms.

The April 2004 issue of *Nature* reports that GM yeast cells synthesise a biochemical substance closely related to artemisinin, a drug proved effective against certain virulent agents of malaria. In the past, artemisinin has been isolated from wormwood (*Artesimia absinthium*).

The most antagonistic adversaries of genetic modifications are Europeans and some North Americans, people whose lives are not at risk from malnutrition or inaccessible health services. Recognising that crop yields from Mendelian plant breeding have levelled off among the developing nations, that throughout Asia there is little additional land suitable for cultivation, that the need for health care, accessible, affordable diagnostics, prophylactics and therapeutics are rapidly growing and diversifying throughout the world, to condemn all genetic modifications as unethical/immoral must be regarded as egotistically irresponsible.

Human Embryonic Stem Cells (HESCs)

In 1998 at Johns Hopkins University HESCs were first isolated from embryos of aborted foetuses. Abortion of a foetus may occur naturally (then frequently described as a miscarriage) or may be induced biochemically or surgically. In the human reproductive process female ova are fertilised by male sperm that immediately form a zygote. Cells of the zygote divide and re-divide at *ca* 20-hour intervals, first into a morula, then to a blastocyst which, some two weeks after fertilisation, is implanted into the wall of the mother's womb as an embryo. From nine weeks after fertilisation it is known as a foetus. Cells that constitute the blastocyst and embryo are pluripotent and can proliferate into any of the roughly 220 different cells that constitute human bodily organs and fluids, e.g., blood cells, brain cells, heart cells.

Many reputable and responsible scientists believe that HESCs could be used to replace or repair dysfunctional body cells that cause chronic morbidity and early mortality through such diseases as Parkinson's and Alzheimer's and through immobilising injuries, for example, to the spinal cord. HESCs can be extracted from very young frozen embryos stored in fertility clinics, embryos left over from *in vitro* fertilisation procedures. Stem cells can be isolated from non-embryonic body tissues but the differentiation potential of most non-embryonic stem cells is largely restricted to the organ tissues in which they naturally reside. They are not normally pluripotent. Whether stem cells harvested from non-embryonic sources can be coaxed to be pluripotent is yet to be demonstrated. Therapeutic cloning consists of replacing the nucleus of a female ovum with DNA from a person suffering from a debilitating disease attributable to dysfunctional organ cells. Early in its development, when the embryo has divided into about 100 cells it becomes the source of HESCs to be used to replace or repair the malfunctioning cells in the sick DNA donor. Since the DNA comes from the sufferer, the implanted stem cells are less likely to be rejected.

Extraction of HESCs can destroy the embryo of source. It has been reported that simple active stem cells can be extraced without destroying the embryo. Whether these single cells are consistently pluripotent remains to be demonstrated.

UK scientists claim to have generated active stem cells, containing 99% per cent human DNA by inserting human DNA into bovine and rabbit embryos. Other scientists report isolation of active stem cells from ambiotic fluid and placentas. Neither of these processes would, it is claimed, destroy the embryo of source.

Certain abnormal stem cells that renew themselves indefinitely actively stimulate proliferation of cancerous cells in human brains, bones and prostrate glands. Radiation and chemotherapies that attack only cancerous cells and leave aberrant stem cells unharmed will not prevent cancer recurrence.

Objections to HESC research arise more from ideological and religious motivations than from scientific considerations. The objectors, who describe themselves as 'pro-lifers', contend that even the youngest embryos are living human beings and to destroy them is an act of murder. Some express reasonable concern that pregnancies and early abortions might be deliberately induced in order to provide HESCs, and that very poor women might be paid to become biological factories. Pro-lifers insist that life begins at the time of conception long before even the most rudimentary nervous system has started to form. Medical scientists however believe life begins much later. It is not very long since one of the religions which now protests that the youngest embryos are living human beings stated that life begins when the mother feels the foetus to move in her womb. Among pro-lifers in the United States there is a curious doctrinal inconsistency. While insisting that extracting HESCs from young embryos is tantamount to murder, they strongly favour the proliferation of firearms and support for ferocious military intervention in countries of whose political regimes they disapprove.

In 2001 the US government forbade the investment of federal money in HESC research. Under Proposition 71 the State of California earmarked $3 million for HESC research. New Jersey, Connecticut and Maryland have also allocated state funds for HESC research. Democratic

Party supporters of HESC state that 'the theological opinions of a few should not be allowed to forestall the health and well-being of sufferers who might be cured by treatment with stem cells'.

A thoughtful article from the University of Berkeley raises pertinent questions:

  i. Who, in each active US State, will determine priorities for HESC research? Should research be long-term focussing on the complexities of stem cell differentiation, or be devoted to immediate pursuit of therapeutic stem cell application?
  ii. Who will own the Intellectual Property Rights to HESC innovations financed by State funds?
  iii. Who will define and enforce regulations related to safety; and whether donors of healthy human stem cells are expected to give altruistically or be compensated for their donations?

In absence of federal government interest, individual State will determine these and other regulatory issues, with protocols likely to differ among States. A comparable confusion exists with the EU where member countries have declared markedly variant attitudes to HESC research.

Functional genomics

This is a field of molecular biology that maps genomes and determines the function and interactions of genes. It enables introduction into experimental animals of gene sequences that encode for cancer and other human diseases. A group of Danish scientists propose three categories of animals used in clinical trials:

  i. Human animals: normal human beings;
  ii. Non-human animals: species such as rodents and primates that may be used for medical biological and other biological research;
  iii. Semi-human animals: animals into whose genomes human genes have been transplanted.

The Danes question if it is ethical to create semi-human animals, indeed should laboratory animals be subjected to possible painful experiment to alleviate human suffering? The World Medical

Association's Helsinki Declaration defines ethical research goals as 'to promote respect for all human beings (Article 8)' Article 11 states that 'where appropriate, experimentation should be assigned to animals'. This deontological, utilitarian concept proposes that human animals are endowed with a privileged dignity denied to experimental animals. To this contentious conundrum there is no logical ethical answer since opinions, pro and con, persist and will continue to persist.

In his book *Rocks of Ages: Science and religion in the fullness of life* Stephen Jay Gould rationally contends that, since their purposes and philosophies are distinct and discrete, science and religion should never be in conflict. In relation to HESCs the conflict between scientists and those of extreme religious persuasion seems irreconcilable. In a nation that claims constitutional separation of church and state, the US Republican administration, strongly influenced by religious interests, has severely restricted HESC research to a few existing stem lines most of which have now degenerated to a state of uselessness. Among members of the European Union relevant regulations range from relatively permissive in Britain to total bans in some other jurisdictions.

Issues anent what is religiously ethical in opposition to what may be medically beneficial are likely to be more pervasive and disturbing as bioscience and biotechnologies advance. Scientists are generally more comfortable when communicating with other scientists than with the general public. If we are to be spared a contemporary parallel with the seventeenth century Galileo imbroglio, bioscientists, biotechnologists, the academic and commercial organisations that employ them must learn how to communicate and enter into more constructive, sensitive dialogue with civil society.

Ethics and sustainable bio-industries

Fifty years ago business ethics was based on three interdictions: don't lie, don't cheat, don't try to deceive customers or shareholders. The late Milton Friedman at the University of Chicago contended that enlightened self-interest should govern ethical business policies; it is most ethical to be profitable since free enterprise is democratic and bankruptcies cause unemployment. To be sustainable every business enterprise must be profitable: over a definable period a company's income must exceed its expenditures; the duration of the critical period will depend on the size of the company, its resources, access to investment capital, loans and credit. Sustainable survival is equally dependent on a clear vision of the company's purposes – the goods and services it offers; its market opportunities – who are its actual and potential customers, what will they buy, in what quantities and at what price; the efficiency with which its resources are managed.

Pharmaceutical and food industries are composed of relatively few, very large corporations that trade internationally, and many smaller entities. The giants grow bigger through acquisitions and mergers; the smaller biotech/biopharm companies survive by offering specialised goods and services to large corporations and other providers of health services. Though they began as small enterprises, some biotech/biopharm companies have grown to substantial size. The largest, Amgen, brings in revenues in excess of $11billion, roughly 20 per cent of which is invested in research and development. Surveys suggest a high rate of failure among small biotech/biopharm companies, many spun off from universities. In several instances of failure, competence in bioscience was not allied with sound business management; others began with insufficient working capital to survive over the long research and development years before a profitable product or service can be developed. As the demand for food and drugs among urban communities expands and diversifies, as costs of production and distribution rise, both pharmaceutical and food companies will benefit by sub-contracting some activities and services to smaller specialist companies. Food processors, particularly those in tropical ecologies, will recognise the advantages of sub-contracting primary processing and preservation of perishable raw materials to rural companies located close to sites of cultivation and harvest.

In all biotechnology industries, practices at one time motivated by ethics or public relations are now required by law. In several countries food processors must list ingredients, caloric and nutrient contents on the package label. Pharmaceutical packages must state recommended doses, active ingredients and warnings to those who might be adversely affected.

Ethics in business

Business ethics is itself big business. The American Business Ethics Officers' Association, founded in 1992, has close to 900 members. Expensive consultants advise companies on their ethical codes and practices: on responsibilities to shareholders, employees, customers and governments; on honesty in trading, advertising, financial management, accounting, price fixing, environmental protection, waste disposal, recycling of packaging, confidential information about customers and employees, equal employment opportunities, racial and sexual discrimination to mention a random selection. Despite the widespread emphasis on adherence to ethical practices, several deplorable examples of conspiracies between managers and accountant auditors to falsify financial records have been publicised. The most egregious having been criminally prosecuted illustrates the wisdom of the 'don't lie, don't cheat' interdictions. Enron has been publicised as the most notorious of offenders, but it seems probable they will not be the last to be indicted. Particularly worrisome are the vast and dilating disparities between a. salaries and other financial benefits bestowed on company CEOs and other senior executives, and b. the wages paid to factory floor workers in the same corporation. It is estimated that total emoluments – salaries, stock options and other benefits – awarded to CEOs of large American corporations amount to 250 times the wages paid to individual factory workers. Similar inequities are evident among industries in rapidly expanding economies such as in India where highly paid managers of information technology companies become progressively richer while rural folk and large urban slum dwellers sink deeper into poverty. Supermarkets pay their CEOs seven figure salaries while employing cheap often part-time immigrant labour in their stores. The cheap garments that Walmart and their like sell are produced by poor Bengladeshi women who work 70–80 hours each week for barely $1.00 per day.

Ethics in international trade

'Globalisation' is a noun often misused to denote international trade. International trade follows many patterns: trade in agricultural and biological commodities between governments and industries; trade in processed foods, drugs and textiles; ownership and control of companies in foreign territories by transnational corporations; inter-company and international licensing of patented technologies and brand names. Three ethical issues related to biological materials and biotechnological industries are of particular significance: (a) subsidised exports; (b) biopiracy; (c) centralised versus decentralised management of foreign owned subsidiaries.

Export subsidies

Ethical issues focus on inequitable access to international markets between poor and affluent nations; on constraints imposed by massive subsidies, by tariff and non-tariff barriers. The World Trade Organisation reports that subsidies to agricultural exports among OECD nations amount to over $350 billion USD per year. These trade distortions permit North American and European nations to export commodities at prices below the costs of production. The WTO, made up of some 134 nations, came into existence in 1995 to administer rule-based procedures for settling trade disputes between and among nations. OECD export subsidies adversely affect many poor African countries. Earnings from primary commodities represent *ca* 40 per cent of Africa's GDP; for 20 nations a single agricultural commodity accounts for more than 50 per cent of export revenues. The World Bank estimates that Africans' share of international non-oil exports is less than one-half their share during the early 1980s.

A praiseworthy movement encourages fair trade in its advocacy that a greater share of profits realised by processors and vendors of coffee, for example, be returned to the smallholders who grow and harvest the coffee cherries and their constituent 'beans'.

Affluent nations protect both their farmers and their processing industries. In North America the tariff rate for raw tobacco is roughly 25 per cent but rises to more than 100 per cent for processed tobacco products. The EU charges average tariffs of 21 per cent for fresh fruits, which rise to 37 per cent for fruit juices. The result is that African economies remain as they did when they were colonies: as suppliers of raw materials with restricted opportunity to develop processing industries. There is some hope that WTO's proposed reductions in subsidies to agricultural exports may become effective with benefit to poorer nations. Given the political influence of agricultural and industrial lobbies in the United States and Europe, it seems unlikely that subsidy and tariff reductions will be substantial, sudden and sustained.

Bio-piracy

Bio-piracy includes exploration, extraction and exploitation of biodiversity from foreign territories for commercial biotechnological purposes. American, European and Japanese agribusiness and pharmaceutical companies are known to collect potentially useful germ-plasm and medicinal plants from poor tropical countries with inadequate if any reimbursement to the countries of origin. Tropical forests are attractive sources of valuable timber and non-timber products for which the countries of origin receive less than fair payment. Pharmaceutical companies, sometimes in cooperation with academic bodies, support extensive ethno-botanical expeditions to collect plants that may embody useful medicinal properties. It is estimated that more than 10,000 plant species and other organisms have been taken from Amazonian forests for examination elsewhere.

Since ancient times Indians have discovered and developed therapeutics from medicinal plants. Records of Ayurvedic and other traditional medicines written over many centuries in Sanskrit, Tamil and Urdu are being entered into the Traditional Knowledge Digital Library, an electronic data base, maintained in Delhi by the Council of Scientific and Industrial Research. A prime objective is to protect proven traditional medicines from being pirated and patented by foreign pharmaceutical companies. With the possible exception of China, no other nation has maintained so comprehensive a record of traditional medicines as has India.

To discourage international biopiracy of its natural resources, in 2002 the Government of India promulgated the Biodiversity Act to prevent removal without due compensation of any biological resource. Contravention of the Act by Indian or non-Indian organisations is a punishable offence.

Centralised versus decentralised management

Several years ago schools of business management debated how transnational companies should manage their foreign subsidiaries. Michael Porter at Harvard University advocated centralised control, all decisions being made and transmitted from the company's headquarters; Kenichi Ohmai recommended that trans-nationals delegate all operational, marketing and policy decisions to the local managers of their foreign subsidiary companies. Based upon experience with transnationals and their subsidiaries this author firmly believes that decentralised control and decision-making are most ethical and effective. Local managers are *au courant* with local conditions, resources and markets and therefore better able to make decisions than managers located long distances away. It is essential that transnationals who establish foreign subsidiaries behave as sensitive corporate citizens in each country of operation, by delegating operational authority, by creative involvement in the foreign communities' social and charitable activities. In the words of a wise executive of a large transnational corporation: "When we establish operations in a foreign country it is in our enlightened self-interest to behave as sensitive corporate citizens".

Summary and conclusions

The contradictions and conflicts between those who favour and those opposed to genetic modifications and HESC research illustrate that MacIntyre and Russell had reason in stating that what constitutes ethical behaviour is based on opinions often influenced by religious, political and ideological persuasions. Not only is there little hope of an Aristotelian golden mean between conflicting opinions, opposition to GM crops and utilisation of laboratory animals in pharmacological and toxicological studies has been demonstrated in violent and destructive actions. Bioscientists and biotechnologists must prepare for future confrontations with opponents of what they perceive as science interfering with nature, with those who believe that humans, even the sick and suffering, are entitled to no greater rights and considerations than other living organisms and early embryos. The contradictions were framed by Benjamin Disraeli's rhetorical question and response: "Is man an ape or an angel? I am on the side of the angels".

Human resource development

For many centuries universities have been organised into faculties or departments, subdivided into specialised disciplines. The disciplinary scope narrows, becomes more concentrated and specialised as students progress from undergraduate to post-graduate studies. All patterns and modalities of development have need of specialist skills and knowledge. The diversity of disciplinary specialisations will expand as environmental conditions and resource requirements change, as new technologies are devised, and to satisfy varying social, economic and industrial needs. International studies of human resources needed by biotechnology industries indicate that present and future demands for bioscientists, biotechnologists and bioengineers will significantly exceed supply over the foreseeable future. Bioengineers capable of expanding unit processes from laboratory and pilot plant to factory scale are internationally in extreme short supply.

Systems analysts and designers

As serious and extensive is the deficiency of systems analysts and designers. A 'system' may be defined as an assemblage of interdependent, interactive entities and activities so integrated as to form a complex unity. All processes of development: social, economic, technological and industrial depend on the systematic integration of diverse resources, resources that must be employed economically and efficiently. This text is not intended as a manual of systems analysis and design methodology. Among authoritative sources of systems methodologies the International Institute for Applied Systems Analysis (IIASA) located in Laxenburg, Austria, has a reputation for exceptional reliability and competence. Suffice it to say that systems analysis may be conceived as a research strategy that enables development managers to select the most efficient course of action and resource utilisation among various alternatives. Systems analysis may be applied to improve an existing system or process, or to design desirable novel systems or processes. Systems analysis begins by critical analysis and assessment of what already exists, what modifications are desirable and feasible, resources to be integrated to realise objective opportunities.

As described in the earlier text, farming systems research is essential before deciding with what tolerable risk novel crop genotypes or production technologies may be adapted. Scale-up of industrial biotechnologies demands systematic analysis of all the interactive and interdependent parameters and variables critical to an economic, technically efficient factory process. Simulation modelling is a mathematical method by which alternative systems of integration can be critically compared, by which a conceptual illustrative diagram of a development process can be created. Simulation models can contrast alternative integrated food systems and compare among different transportation and distribution opportunities, locations for storage depots, primary and secondary processing facilities, which will be most logistically efficient and economic.

Physiological functions depend on interaction among many genes and the environments in which proteins are synthesised and function. Simulation modelling provides more reliable comprehension of cellular, tissue and organ functions. The International Union of Physiological Sciences encourages definition of common standards, of a language and vocabulary to permit sensible precise communication among computational modellers. A simulation model is designated 'deterministic' when based on quantitative data, as 'stochastic' when past factual experience is merely indicative of a probable outcome (Greek: 'στοχαστικος' = 'guess' or 'conjecture'). As development processes become increasingly complex, systems analysts and mathematical simulation modelers will be increasingly in demand by planners and industries.

Simulation models, designed to compare the relative utility and/or economy of alternative complex system require the engagement of highly specialised mathematicians simulation models, which may be displayed as computer graphics defined on probability mathematics, partial differentials and designed algorithms. 'Algorithm' is the Latin name of an eminent 9th century Persian mathematician al-Khawarazmi whose algorithm prescribed arithmetic rules based on Arabic numerals. Algorithms are a mathematical means of controlling sequences in computer programmes. Complex algorithms permit more precise interpretations of mammograms and other diagnostic images that detect cancerous cells.

As Pearson, Brandt and Brundtland discussed, many past development programmes intended to eradicate poverty, to improve economic, social and technological conditions failed because of an over-simplistic perception of the complexities entailed. Development, for whatever purpose, must begin by comprehending and systematically analysing the environments and conditions that exist among the communities to be assisted. Systems analysts, invariably necessary, have been all too rare in past development programmes.

Language and communications

This closing section brings us practically full circle to the issues of language and meaning. In his publication *Of Studies* Francis Bacon wrote: "Reading maketh a full man; writing an exact man." Would that it were so! As illustrated throughout the foregoing text the language in many publications from development administrators and scientists is neither exact nor precise. There is no shortage of scholarly, instructive books that offer cogent advice to writers, including the classic by Arthur Quiller-Couch [1916, 1938], by Fowler [1926, 1930], Gowers [1954], Rama [1950], Russell [1957] and Cooper [1971]. Though their contents differ in substance and dimension, the authors listed are unanimous in their insistence that those who write should be precise and concise; should not use words they cannot define, words they do not fully understand; and should not coin or invent words that have no generally accepted definable meaning (e.g., 'sustainability').

Words, particularly from the pens of poets, can have several meanings: colourful, metaphorical, allegorical or persuasive. The noun 'quality' is derived from the Latin *'qualitatis'* which means literally 'of a particular kind': the combined properties of a substance, material or product which distinguish it from all others. 'Quality' is used by advertisers to suggest superiority: a 'quality' product is presented as better than all its competitors. At one time, tailors, grocers and other British vendors identified 'the Quality' with their more affluent customers.

The literature related to 'Sustainable development' is sated with imprecise, undefined vocabulary. 'Development' has no determinable meaning unless the specific objectives and the criteria by which progress in the development process will be measured or assessed are precisely defined. In proposed 'sustainable development' programmes it is frequently unclear whether it is the development process or the objectives of the process that are to be sustained. In either case it is essential to define who or what are to be sustainably developed, with what specific end objectives, with what resources and over what period of time – for months, years or indefinitely.

Evolution of the means of communication

It can be argued that of all recorded ingenuities most beneficial to humanity were discovery of the wheel and the invention of writing, both of which came from Mesopotamia many centuries ago. Writing in cuneiform, well established by the fourth millennium BCE in the ancient Mesopotamian city of Uruk, took the form of a complex assortment of wedge shapes (Latin *cuneius* = wedge) imprinted into damp clay by a stylus fashioned from a reed.

Cuneiform writing was invented for practical purposes. As cities, agriculture and agribusiness expanded and activities diversified, efficient administration had need of reliable records. The earliest cuneiform tablets discovered record activities in textiles, pottery, crops and livestock and in trade between producers and customers. Each city had its patron god and the cuneiform records were to keep accurate accounts of the god's household – Greek: οικος (oikos) from which οικονομεια (oikonomeia) 'management of the household' (in English 'economics') is derived. Cuneiform inscribed both words and numbers: > = 10, >> = 20, >ll = 12 (where the two letters 'l' would appear as long, narrow triangles).

Sometime during the Pharaonic era, Egyptian priests devised a sacred language printed in hieroglyphics (Greek ιερος = sacred, λυφη = carving). Hieroglyphs were inscribed on sheets of papyrus, constructed by pressing lightly together thin films shaved from the stems of an aquatic plant *Papyrus antigonim*, a species of the sedge family, which includes the lotus blossom. The pressed thin films were then dried. The port of Biblos from which papyrus scrolls were shipped to Greece gave its name to 'the scrolls' or 'the books' (βιβλια) which constitute the Christian bible.

The earliest various systematic collection of printed knowledge existed in libraries composed of cuneiform on dried clay tablets, and hieroglyphs printed on scrolls of papyrus. Many scholars, including Aristotle, amassed their own personal libraries. Some opened their collections to other scholars who enhanced their knowledge by consulting the accessions in the collections. The Assyrian King Assurbanipal

(668-627 BCE) was one of the most zealous archival collectors and is believed to have amassed a library of several thousand accessions.

The most remarkable and extensive of all early libraries was the Bibliotheca Alexandrina (Greek: βιβλιον [biblion] = scroll or book; θηκη [theke] = depository). When Alexander the Great, Aristotle's most renowned pupil, founded the city of Alexandria, he commanded that there should be a library and museum dedicated to the muses. After he died in 323 BCE his Macedonian successors, who became Kings of Egypt – respectively Soter, who became Ptolemy I, his son Philadelphus, Ptolemy II and grandson Eurgetes, Ptolemy III – fulfilled his wishes and established the Bibliotheca. It was Alexander's intention that the Bibliotheca be repository for all the world's publications: printed scrolls and incised clay tablets, and that it be a centre where scholars would meet to exchange and advance the sum of human knowledge. Archimedes and Euclid were two of the many distinguished scholars who spent time at the Bibliotheca. The Ptolemy dynasty continued until Ptolemy XIV when Marcus Aurelius and Cleopatra, sister of Ptolemy XIV, were defeated by Octavius (later to be Augustus Caesar, first Emperor of Imperial Rome) at the battle of Actium in 31 BCE.

It is immensely pleasing and a tribute to the Government of Egypt that a new Bibliotheca was officially opened in Alexandria in 1995. Its stated purpose is to be a home for much of the world's scholarly literature and a well equipped centre which will attract scholars from around the world. The Bibliotheca embodies a museum of antiquities with an expanding collection of artifacts representative of ancient Egyptian, Grecian and Roman civilisations. A huge modern conference centre with four auditoria is alongside a planetarium, a library equipped to house 80 million printed works, and a computer complex destined to record and classify all writings worthy of retention and dissemination from the Internet and related sources, all of which will be accessible to scholars throughout the world.

Ancient fragile documents have been carefully photocopied and can be viewed on computer **screens so** programmed that pages can be turned by stroking the screen with a finger, as one would turn the page of a printed book. Thus precious, fragile documents are protected from damage by the hands of readers.

A remarkable virtual reality laboratory projects in three dimensions, to spectators wearing special spectacles, molecular structures of substances believed to have been among the first when the cosmos came into existence. The molecules and other projections can be moved around, with additional features, such as electrons around atomic nuclei, being added by hand-held controls.

Mechanical printing

The earliest printing press was devised in Mainz in 1449 by Johann Gutenberg who invented a type-casting machine with moveable metal type cast in individual letters. Printing has subsequently progressed to systems whereby book texts and illustrations can be transmitted electronically from author to publisher, and printed from diskette or CD ROMs.

Transmission of information

The first electromagnetic telegraph was invented in 1832. In 1836, a Russian Pavel Shilling improved the apparatus and transmitted signals across 5 km via an underground-underwater cable in St Petersburg. In 1892 Thomas Edison, who was deaf from early childhood and worked as a mail clerk on an American railroad, patented a two-way telegraph.

The Morse Code was intended for Samuel Morse's electrical telegraph, invented in 1837. The Code was devised jointly by Morse and his colleague Alfred Vall and was most extensively used when Marconi's radio communication invention became available in the 1890s. The Morse Code, composed of various combinations of dots and dashes, can transmit both letters and numbers as electrical pulses, audio tones and visual signals. The best known are ... _ _ _ ... (SOS). The number one is transmitted as one dot followed by three dashes. Morse Code is now largely obsolete, except for emergencies when no other means of communication are available.

Reuters

Reuters news service, which provides reports of current events to newspapers, broadcasters and financial markets around the world, was founded during the 1860s by a German immigrant to Britain, Paul Julius Reuter. At first, Reuters transmitted news via the Calais-Dover telegraph cable and across Europe by carrier pigeon. It now employs several thousand journalists and devotes most of its resources to collecting, analysing and transmitting financial information and statistics. In 1984 it became a public company, registered on the London Stock Exchange.

Telephone

The name derives from two Greek words: 'τηλε' that means 'distance' and 'φωνη' translated as 'voice'. Who invented the first telephone is still contested among several men. Antonio Meucci demonstrated in 1849 in Havana a sound transmission device involving electrical connections between transmitting and receiving stations. He then demonstrated his electric telephone in 1854 in New York, and again in 1860 in Staten Island. In 1861 a German inventor, Johann Phillip Reis, transmitted voice sounds over a distance of 340 feet. In 1875 Alexander Graham Bell transmitted voices via a two-way directional telephone, the transmitter and receiver being identical membrane instruments. Most North Americans regard Bell as the official inventor of the telephone. In 1877 Emile Berliner demonstrated his invention of the microphone.

During the past century, particularly since World War II, extensive and diverse developments in telephonic technologies have appeared, many attributable to the Bell Research laboratories in the United States and to research scientists employed by the British postal service. Since ownership of telephonic companies and services was deregulated, many companies have entered the field, several combining telephone services with Internet and other electronic means of communication.

Neglect of development progress

Despite the ready availability of means of instant voluminous communications, many development agencies fail to maintain a reliable corporate memory or a lively cognisance of earlier mistakes and successes, or of activities by other agencies worthy of their cooperation.

International development agencies notoriously display little interest in what becomes of the products of their aid assistance after the aid investment comes to an end. Many are instances in which a poor nation, institution or community has been saddled by an aid agency with a facility and/or equipment they cannot afford to maintain or sustain; machines and instruments that need spare parts available only for foreign currency; technologies transferred from the donor nation ill-suited to the recipient's physical, social and economic environments and resources.

As was stated on the subject of systems analysis, all development programmes and projects must start with a comprehensive, critical understanding and appraisal of what already exists before any development programme or process is planned and set in motion. Included must be a shrewd assessment and systematic comprehension of the human, financial and material resources required to pursue the development objectives, and to sustain what the development process eventually creates and establishes. All development proposals, plans and progress should be recorded in precisely defined language, in terms that all who are initially involved and those who come later can clearly comprehend.

Much is written and spoken about 'globalisation' a noun interpreted differently by different people. To plagiarise the *Meditation* of John Donne: '*Nunc lento sonito dicunt, morieris*'; no nation, corporation or agency is an island, entire of itself, everyone is 'part of the main', a member of the world of nations and interdependent communities.

International trade, international mergers and alliances are of increasing frequency. Groups of nations ratify agreements to facilitate trade among member countries. Electronic communications permit rapid exchange of knowledge and information among scientists, corporations and governments. To communicate intelligibly there must be a common comprehensible vocabulary. The difficulties of creating and maintaining intelligible bioinformatics networks are illustrative of a worldwide challenge to scientists and semasiologists. Instead of more international mega-conferences where delegates regurgitate the time-worn weary platitudes about development and the environment, urgently needed are professionally competent specialist working groups, each to include philological expertise, to bring into being a series of international thesauri, each to serve a specific specialised field of scientific, economic and industrial research and development. The International Union of Physiological Sciences sets an example worthy of adaptation by other agencies and scientific organisations.

Samuel Johnson wrote that knowledge is of two kinds: either we know a subject ourselves or we know from where we can obtain relevant information. Relevant knowledge and information are critical to all modalities of development. The world's complex development needs are beyond any one nation's or agency's capacity to provide. International cooperation that brings into productive cooperation creative intellects and resources wherever they exist are essential to sustainable human survival. Design, development and propagation of intelligible vocabularies could provide the most valuable assets to all who pursue sustainable development for the whole of humanity. As the remarkable scientist to whom this book is dedicated so often stated, science and scientific knowledge serve no useful purpose if they do not serve the needs of humanity, especially those who are in greatest need.

Epilogue

In his 1932 analytical history of the human race *Adventure of Ideas*, the distinguished mathematician and philosopher Alfred North Whitehead writes that his book's contents are "dictated by the arbitrary limitations of my own knowledge". Every historical text is inevitably constrained and conditioned by the author's knowledge, his/her access to reliable information, how he/she interprets the information considered. Socrates claimed he knew nothing save his own ignorance. During the early fifteenth century Nicolas da Cusa observed that "the absolute truth is always beyond our grasp". This publication is a mixed compilation of verifiable facts, extracts and summaries from many sources, together with comments, observations and recommendations inevitably coloured by the author's beliefs, opinions and interpretations.

The foregoing text explores and seeks to explain diverse concepts of development and 'sustainable development', a popular term frequently written and spoken with little precise definition of what it is intended to mean. While many social and biological systems remained relatively unchanged over many centuries, modern developments – biotechnological, social, economic, industrial, informational – are changing rapidly. Most such developments are controlled by complex interactions of many variables. Few are susceptible to simple solutions and demand the skills and experience of men and women highly proficient in systems diagnosis and design. Developments progress most effectively when guided by the dictionary definition of 'development': "a gradual (and systematic) unfolding".

No longer can development be measured simply by growth in wealth and prosperity, among national economies by increase in GDP, among commercial companies by rising share prices quoted by stock markets. Development, both national and commercial, entails consideration of human values. It is incumbent upon all governments to ensure equal rights with access to an equitable and secure standard of living for all its citizens; the most valuable assets of commercial enterprises are its employees and its customers. It is shameful that senior executives in North American companies receive salaries and financial rewards many times greater than the wages paid to workers on the factory floor, and the even worse wages paid to poor immigrants casually employed on farms and in sweatshops.

Substantial space is allocated to International Development: the provision of technical, financial and other assistance by international, national and non-governmental agencies to poorer countries and communities. Bilateral assistance programmes provided by affluent nations have been inconsistent, volatile and retrogressively inhibited by an addiction to tied aid, designed more for the benefit of the donor's farmers, industries and overall economies than in determining and responding sensitively to the urgent needs of the recipient. National bilateral agencies seem more intent on competing than on cooperating in long-term commitments to systematically designed development programmes. Though alleviation of poverty has been a stated development objective for more than 50 years, both within and among nations, the rich become richer as the poor become poorer.

While applauding the spectacular economic and industrial growth in China and India, one cannot but be depressed by the disproportionate distribution of the wealth amassed by the rich minority and the impoverished majority in both nations. It would seem that those in authority in India have forgotten the benign teachings of Mahatma Gandhi, and that the Chinese government has ceased to be guided by its professed socialistic principles.

The changes in policies that come about as governments and directors of development agencies change are the most obvious deterrent to consistent commitment. Governments and senior executives seem intent on enforcing their policy priorities even though worthwhile programmes are discontinued in consequence. At the time of writing, ODA flows to the poorest nations are depressed by diversion of aid funds to finance reconstruction of properties and services in Iraq and Afghanistan, many destroyed by the aggressive bombardments of foreign invaders. Support for the so-called war on terror takes priority over alleviation of poverty both within and among nations.

Many governments, their bureaucracies and advisers appear woefully unaware of the debilitations that result from overly rapid, unplanned urban expansion and congestion. Proliferation of slum conglomerations with unhygienic housing and sanitation, inefficient waste disposal, inadequate access to clean potable water, overcrowding conducive to the spread of pathogenic infections were stimulated by the mechanisation of agriculture in the United States and the industrial revolution in Britain during the eighteenth and nineteenth centuries. Comparable but even worse conditions are evident in uncontrolled, unplanned fast growing cities and large towns across Africa, Asia, the Middle East and Latin America.

From the earliest historical records of ancient civilisations, religious leaders have admonished the rich and powerful to act with beneficence and to protect the poor and vulnerable members of their societies. It is wickedly unethical for any civilised community or religion to claim a divinely endowed right to steal by force the land and property of weaker nations, or to seek to impose their beliefs and doctrines by means of military might on others of different persuasion. It is not conducive to peaceful relations for any extremist movement to proclaim possession of a divinely bestowed right to propagate what they accept as divinely ordained doctrines. Aggressive religious bigotry has inflicted so much pain and suffering on innocent people in the past that one might expect civilised human beings, particularly those whose ancestors have suffered extreme persecution, to be sympathetic and tolerant of those who profess beliefs and ideologies different from their own. No nation or ethnicity has the right to claim divinely endowed superiority over all other peoples.

It is indeed difficult to be optimistic that all living and future humans and other earthly creatures can be assured of survival and a healthy, sustainable secure future. The planet's atmosphere is being poisonously polluted by emissions from the combustion of fossil fuels; increasing contamination by greenhouse gases threaten irreversible global warming and climate change seriously detrimental to the stability of many ecosystems. Arctic ice is melting at unprecedented rates, foreboding devastating rises in sea levels that could cause drastic inundations of coastal regions which, as demonstrated by the damage wrought by Hurricane Katrina, even the world's most affluent nation is powerless to prevent.

Both coastal waters and deep oceans have been excessively over-fished. The 3 million fishing vessels that regularly put out to sea threaten the extinction of more than a quarter of the surviving ocean fish stocks. Japanese, Korean and other modern fishing fleets use sonar to locate vulnerable species as well as drag nets of over 50km in length to scoop up all aquatic life that crosses their path. These nets harvest and destroy both adult and juveniles, together with commercially unwanted species, thus irreparably devastating aquatic populations. It is discouraging when nations who profess dedication to environmental and biodiversity conservation act in direct contradistinction. Japan, home to the Kyoto accord, and Norway, whose former prime minister Mme. Brundtland instructed all nations to act so as not to put at risk the needs of future generations, now use every means at their disposal to reverse restrictions on commercial whaling, the destruction of one of the most endearing and endangered aquatic creatures. Neither Japan nor Norway can claim to be deficient in alternative sources of animal protein.

Pollution by ships, oil spillage, effluents from coastal industries and communities, together with deep sea drilling, add to the toxic aqueous environment in which many aquatic species and coral reefs are destroyed. A UN committee reports that during 2005 more than 250,000 turtles and 300,000 sea birds were killed by entanglement in fishing nets. Over one million birds perished from contact with garbage discarded into the oceans. Global warming, melting of Arctic ice, pollutions of oceans, coastal and inland waters, extensive deforestation collectively portend unremitting depletion of biodiversity, the extinction of countless organisms, many valuable to agriculture or medicine, others, great and small that enhance the variegated beauty of our world.

Land and ground water is poisoned by the disintegration of over 15 million cell phones and unknown numbers of personal computers (the average life of which is a little over three years) discarded into garbage every year. The vandalising, irreversible waste of natural resources by greedy nations, who already possess more than their fair share of the worlds goods, sets an intolerable example which, if emulated by industrially emerging nations, will inflict widespread suffering and threaten the survival of millions of poor people by making life on earth wholly unsustainable.

While powerful governments use their military might to grab control of the fossil fuels of weaker nations, it is access to clean water that will determine the future survival of many peoples. Of all resources essential to human survival, fresh water is most critically in short supply. Agriculture, industries, municipalities and countless communities compete for scarce water. Much water is wasted by mismanaged irrigation, by leakages during transmission, by deliberate and adventitious pollution, by inadequate purification and recycling, by the failure of governments to charge users for the true cost of the water they consume. Rivers, lakes and other surface waters are widely regarded as nobody's and therefore everyone's property, free to be drawn on or polluted by industries, farmers and the general public. Competition for access to restricted water among nations is more likely to result in future armed conflict than the pursuit of fossil fuels.

Among biotechnologically advanced nations during the past century, the progressive transition from artisanal empiricism to scientific innovation and control has been impressive. A more precise understanding of molecular and cellular biochemistry, of genetic structures and functional modalities, has stimulated immense progress in agriculture and medicine. Cell cultures of genetically modified organisms, of microbial, insect and mammalian tissues synthesise novel diagnostic, therapeutic and prophylactic drugs. Productive research that discovers the influence of genetic disorders and bio-molecular interactions on chronic diseases and malignancies, is paving the way for future medical concentration on disease prevention.

Under some jurisdictions progress in agriculture and medicine is impeded by ideological opposition to all genetic modifications, to the potential use of pluripotent embryonic cells to repair or replace dysfunctional body tissues in patients who suffer from various degenerative diseases. There is little justification for religious or other ideologues who enjoy a healthy, wealthy life to deny alleviation of chronic degenerative diseases among people so afflicted.

Given the increasing longevity among many populations and the higher susceptibility to degenerative and other diseases among older people, added to the rising costs of diagnostic devices, drug development and health delivery services, it is improbable that even the wealthiest nations can afford to provide entirely free total health care for all their citizens. The alternative options of raising the age for compulsory retirement, of increasing taxes and/or personal contributions to health care insurance, demanding partial payment for all health care services are eschewed by politicians unwilling to annoy the constituents they hope will re-elect them.

The final chapter reflects on the remarkable progress, since writing was invented some 5,000 years ago, in systems of communication.

As so many communications are transmitted electronically between organisations and people who never meet face-to-face, there is urgent need for more accurate precision in spoken and written language to ensure that messages transmitted convey an identical meaning to both the transmitter and the receiver. Imprecise language inevitably leads to misunderstanding and misinterpretation.

The changing complexities of modern industries, efficient governance and urban expansion and congestion demand the active attention of men and women with experience in systems diagnosis and design. Many past failures have resulted from reliance on solitary specialisations; from inability to plan and implement developments holistically and systematically. Agriculture, industrial biotechnologies, medical services and treatments, international and countless other development programmes involve interactions among complex component variables, many yet to be discovered. For any development to be sustainable it must be conceived, created and controlled holistically and systematically.

Opportunities offered by modern science and technologies for beneficial human, social and economic development are seriously threatened by the disposition among powerful nations and factions to settle all contentious issues by force of arms. Expenditures on weapons of war far exceed investments in environmental conservation, disease eradication and alleviation of chronic poverty. The review conference of the Nuclear Non-Proliferation Treaty (NPT) held in 2005 in New York ended in deadlock. The five permanent members of the UN Security Council, all possessors of vast nuclear arsenals, were unwilling to eliminate their nuclear weapons. It is these same nations, supposedly responsible as Security Council members for ensuring worldwide peace, who derive considerable income from selling their obsolescent armaments and who dictate which of their friends and allies have the right to own weapons of mass destruction and which nations shall be denied that right. Sustainable development for all humanity and worldwide peace will be realised only when the most powerful nations accept that all peoples have equal rights, that violent confrontations result from gross inequities, from abject poverty and unemployment among young men and women.

Closing comment

All learning, all worthwhile development is derived from a critical awareness and understanding of past experience. As the French philosopher Henri Bergson wrote in his book *Creative Evolution*: "Durable development is the result of continuous progress from the past, which gnaws into the future, [it] swells as it advances … and unfolds gradually". For there to be sustainable development for all humanity and the places where they live, there is urgent need for a new world order, one that ensures equivalent rights and opportunities for all peoples, irrespective of their creed, colour or ethnicity. An equitable world order cannot be exclusively based on Western concepts of governance and ceaseless material acquisition. A benign and equitable vision of international human, social and economic development must sensitively accept a wide diversity of cultural, social and religious dispositions.

Barbara Ward, a member of IDRC's founding Board of Governors, concludes her profound and perceptive 1961 publication *The Rich Nations and the Poor Nations* with the following words, as relevant today as when they were written: "We reap what we sow, and if freedom for us is nothing more than the right to pursue our own self-interest – personal or national – then we can claim no benign vision for our present or future society. Without a benign vision we, like others before us, will most surely perish".

Glossary of Biotechnologies

Abiotic stress: Caused by a non-biological agent eg soil salinity.

Adenine: A purine nucleic acid base in DNA [pairs with thymine] and RNA [pairs with uracil].

Aerobic: Organisms that grow in presence of oxygen.

Agrobacterium tumefaciens: Bacterium that infects plants; contains a plasmid used to transfer genes between organisms.

Amino acids: Nitrogenous chemical compounds; the structural components of proteins. 20 present in natural organisms.

Anabolism: Chemical changes in living organisms to store chemical energy in complex substances.

Anabolite: Substance essential to an end product of anabolism.

Anaerobic: Organisms that grow in absence of oxygen.

Angiosperms: Flowering plants in which seeds develop and mature inside closed ovary.

Anther: Surmounts the stamen [plant's male organ] and carries the pollen sack. Pollen released when anther ripe.

---

* The Glossary does not define or describe artisanal and traditional technologies of preservation and transformation practised for many centuries.

Anther culture: Culture of cells derived from excised anthers.

Antibiotics: Metabolites isolated from microorganisms that inhibit growth of rival organisms. Anti-microbial therapeutics.

Antibiosis: Mutual antagonism between living organisms [Antonym of "symbiosis"].

Antibody: Protein synthesised by immune system to counteract an antigen. Immunity to infection permanent or temporary.

Anticodon: Base sequence in tRNA molecule that recognises a codon on mRNA.

Antigen: Foreign substance on surfaces of pathogens that stimulate immune system to synthesise protective antibodies.

Apical meristem: Tip of new plant shoot or root from which cells may be cultured to generate disease-free plants.

Asexual reproduction: Plant/animal reproduction without fusion of male/female gametes: Vegetative, tissue, cell culture.

ATP [adenine triphosphate]: *In vivo* conversion to mono-diphosphate provides energy for eg respiration, N-fixation.

Attenuation: Pathogenic microorganims deprived of virulence; As vaccines to stimulate antibodies and provide immunity.

Bacteriophages: Viruses that infect bacteria [Used to transfer DNA from unrelated organisms].

Bacterium: Unicellular organism; reproduces by division. Genetic code in bacterial chromosome: tangled coil of DNA.

Baculovirus: Type of virus that infects insects; used as vector to produce foreign proteins in insect cells.

Base: Reacts with acids to form salts; *Purine, pyrimidine bases* cross link in twin helixes of DNA.

Biocontainment: Prevention of pathogens, other microorganisms to escape from laboratory or experimental site.

Biological Oxygen Demand [BOD]: Amount of O_2 to destroy organic matter by aerobic organisms. Index of organic pollution.

Biomass: Total biological material produced by microorganisms, plants and/or animals.

Biopolymers: Macromolecules synthesised by living organisms [eg proteins, carbohydrates].

Biotechnologies: Processes that produce, protect, preserve and/or transform biological materials into products of commercial, economic, social and/or hygienic value and utility.

Callus culture: Mass of undifferentiated cells cultured from an explant. First step in plant generation by cell/tissue culture.

Catabolism: Metabolic processes that liberate energy [cf Anabolism, Metabolism].

Cell culture: Colony of cells propagated from a single cell in nutrient medium.

Cell fusion: Fusion of two or more cells from a single cell.

Cellulose: Linear polymer of glucose units: provides structural framework of plants; walls of some cells.

Chimera: Organism or DNA constructed from two or more species. Human cell nucleus inserted into the egg of another animal species.

Chloroplast: Plastids that contain chlorophyll; photosynthetic 'factories' within leaves.

Chromosome: Linear sequence of genes in cell nuclei; in bacteria as 'naked' DNA; In higher organisms: DNA plus protein.

Chromosomes: Units of transmission between generations. When cells divide may break, form new genetic combinations.

Clone: Genetically identical cells or organisms asexually derived from common ancestor.

Cloned embryo: A human egg into which the progenitors cell nucleus has been inserted.

Coding sequence: Segment of gene expressed to translate into protein.

Codon: Initial letters of 3 nucleotides in sequence in messenger RNA; each triplet for a specific amino acid.

Colchicine: Alkaloid in Autumn Crocus root: biochemically doubles number of chromosomes in haploid cells.

Cotyledon: Embryo leaf in flowering plant. Angiosperms classed as Monocotyledonae [1 Cot] and Dicotyledonae [2 Cots].

Cytoplasm: Semi-fluid that surrounds nucleus [nucleoplasm] in cells.

Cytosine: Pyrimidine nucleic acid base: pairs with guanine in DNA and RNA.

DNA [Deoxyribonucleic acid]: Polymer: carries and transmits genetic code; controls cellular functions in most organisms; twin helical strands composed of deoxyribose, phosphate, adenine, cytosine, guanine, thymine; cross-liked A:T,C:G.

Embryo: Immature organism that results from union of male and female gametes. In plants a *sporophyte.*

Embryonic stem cells: Primitive cells extractable from young embryos. Pluripotent: generate most cells in host organism.

Human embryonic stem cells can generate most human body cells [eg blood, heart,]. Potential to renew defective organs.

Embryo rescue: Generation of fertile plants by culture of embryos from crosses of genetically unrelated plants.

Endonucleases: Enzymes that recognise specific base sequence; break DNA into short strands coded for desirable trait [See Restriction enzymes].

Entomogenous microorganisms: Bacteria that infect insects.

Enzymes: Proteins that catalyse specific biochemical reactions that break down or synthesise biochemical compounds.

Eukaryotes: Organisms with cells that contain nuclei containing chromosomal DNA [cf Prokaryotes].

Exon: Gene segment expressed by translation into a specific protein.

Exonucleases: Nuclease enzymes that attack nucleic acids from the ends and shorten the strand.

Explant: Excised plant tissue to be used for cell/tissue culture.

Fermentation: Biochemical syntheses of metabolites from microorganisms [or enzymes] cultured in nutrient substrate.

Gamete: Mature sex cell able to unite with gamete of opposite sex to create embryo for new plant/animal.

Gene: Linear unit of heredity transmitted between generations by sexual asexual reproduction. Segment of nucleic acid in DNA encoded for a specific protein or RNA molecule.

Gene expression: Chromosomal genes 'switched on' and active. Some genes active others silent in particular organs N B When foreign genes are transferred between organisms they must be activated. More difficult in plants than bacteria.

Gene mapping: Determining location of different genes on a chromosome.

Gene therapy: Replacement of a defective gene in a person suffering from a genetically caused disease.

Genetic code: Relation between triplet nucleic acid codons and the amino acid for which each codon is specific.

Genome: Total genetic complement of an organism; Genetic composition of chromosomes in gamete nucleus.

Genotype: Genetic composition of an organism cf its physical [phenotypic] characteristics.

Germ plasm: Genetic material [eg seed] from which organisms are propagated.

Germ plasm bank: Organised collection of seed, sperm, ova from which plants or animals may be selectively propagated.

Guanine: Purine nucleic acid base which pairs with cytosine in DNA, RNA.

Halophile: An organism tolerant to sodium chloride.

Haploid: A cell that contains half the number of chromosomes in somatic cells. Sex and germ cells are typical.

Helix: A spiral form coiled around a central axis. Eg a corkscrew.

Heterokaryon: A cell in which two or more genetically distinct haploid nuclei coexist and multiply [synonym: Polynucleate].

Heterosis: Significant difference in the mean value of a character in an offspring cf the mean value in the parents.

Heterozygous: An offspring that for any specific character inherits different genes from its parents.

HIV: Human immunodeficiency virus: a retrovirus.

Homozygous: An offspring that for any specific character inherits identical genes from both parents: therefore produces genetically stable gametes.

Hormone: A chemical messenger secreted by endocrine or ductless glands carried by blood to induce a specific response in a target organ. [adrenaline stimulates heart functions; insulin controls carbohydrate metabolism].

Hybrid: An organism derived from crossing two different genomes [From sexually compatible parents or by transgenesis].

Hybridoma: Hybrid cell derived from fusion of a cancer cell and a normal cell. If from spleen lymphocyte cell, fused cells may be cloned to secrete pure antibodies [cf Monoclonal antibodies].

Insulin: Hormone synthesised in animal pancreas that controls glucose metabolism. Diabetics suffer from insulin deficiency. First natural protein to be synthesised in vitro. Now synthesised in genetically modified bacteria.

Interferons: Proteins that appear in blood of mammals in response to viral infections.

Intergeneric hybrids: Hybrids of two species each of a different genus [eg Triticale wheat [Triticum] x rye [Secale].

Introns: Nucleotide sequences in DNA, interspersed with exons, not expressed by protein synthesis.

In vitro: Literally 'in glass'. Natural biological processes experimentally reproduced outside a living organism.

In vivo: Biological processes within a living organism.

Karyon: A cell nucleus. Eukaryon – complex cell nucleus of higher organisms; Prokaryon – nucleus of microorganism.

LD 50: The dose of a toxin or pathogen that will kill one half of a population of organisms.

Leucocyte: White blood cell.

Ligase: Enzyme that catalyses union of two molecules. Used to join strands of DNA in recombinant techniques.

Lignins: Complex phenolic plant polymers of variable composition.

Lignocelluloses: Polymers composed of lignin and cellulose essential to structures of woody plants.

Lymphocyte: Leucocytes synthesised in lymph glands and spleen that stimulate antibody synthesis *in vivo*.

Lysis: Physical, chemical, enzymatic degradation of cell walls to release contents.

Meiosis: Division of sexual cells each nucleus containing half the chromosomes in normal somatic cell. When sexual cells fuse, resulting embryos contain full chromosome complement, half from each parent cell.

Meristem: Tip of new plant shoot or root that produce rapid proliferation of cells.

Meristem culture: Cell cultures from shoot of root tip of a plant. Can generate disease free plants from infected parent.

Messenger RNA: mRNA conveys genetic code from the DNA to ribosomes where code is read and protein synthesised.

Metabolism: Total sum of biochemical, biophysical changes that occur in a living organism.

Metabolite: Biochemical product of a specific metabolic reaction.

Milt: The sperm of male fish.

Mitochondria: Filamentous protoplasmic bodies in cell cytoplasm that carry enzymes to catalyse synthesis of compounds that store energy.

Mitosis: Somatic cell division [non-sexual] by which each daughter cell contains chromosomes equal to parent cell.

Monoclonal antibody: Pure antibody derived from a single clone of hybridoma cells that recognises only one antigenic site. Hybridoma cells may be produced by fusing an antibody producing cell [eg, a lymphocyte] with a tumor cell. Because of its specificity, each MBA applicable to a specific diagnostic or therapeutic purpose.

Morphogenesis: Regeneration of whole organism or organ from sexual or somatic cells.

Mutation: An inheritable change in an organism's identifiable properties caused by alteration of one or more genes.

Mutagen: Chemical or physical agent or process that induces mutation

Mycorrhiza: A symbiotic relation between a fungus and the roots of a higher plant that increases the plant's capacity to absorb soil nutrients [eg, phosphates].

Nitrogen fixation: Conversion of atmospheric nitrogen to a form that can be utilised by plants.

Nuclease: Enzyme that catalyses hydrolysis of nucleic acids.

Nucleic acid: A sequential chain of sugars, phosphates, a base attached to each sugar, which constitute the genetic code.

Nucleo-proteins: Nucleic acids [DNA] combined with protein molecules in cell nuclei; main constituents of viruses.

Nucleotide: Sugar molecule with attached phosphate and base molecules that link through phosphate to from nucleic acids.

Oncogen: Substance or organism that stimulates cancer cell synthesis

Oncogene: An activated modified gene that causes normal cells to become cancerous.

Oocyte: A cell that divides to form a female reproductive cell.

Organelles: Protoplasmic bodies in cells with specific functions [eg, nuclei, mitochondria, ribosomes].

Pathogen: An organism that infects to cause disease.

Peptides: Chemical linkages by which amino acids are joined to form protein polypeptides.

Phenotype: Physical characteristics [eg, size, morphology] of an organism resulting from interaction of genetic expression with surrounding environment.

Phytopathogen: Organism pathogenic to one or more plant species.

Plasmid: Small circular DNA able to reproduce independently of main chromosomes. Foreign DNA sliced into plasmids and transferred between microbial cells by bacteriophage vectors whereby used as means for genetic transformations.

Ploidy: The number of sets of chromosomes present in an organism or cell.

Pollen: Powdery material contained in sacs on plant anthers. Each pollen grain carries two male nuclei: two male gametes.

Pollen culture. Culture of pollen cells to generate progeny each with single set of chromosomes [haploids].

Polymer: Macromolecular substance composed of smaller molecules bonded together in chains which may be cross-linked.

Polymerase chain reaction [PCR]: A method of generating quantities of a target segment of DNA.

Prokaryotes: Simple organisms [eg bacteria] without a distinct nucleus. Heredity controlled by a naked strand of DNA.

Protoplasm: Semi-fluid complex material of which all living organisms are composed.

Protoplast: Plant cell from which cell walls removed by physical or enzymatic means.

Protoplast fusion: Union of two protoplasts to form a hybrid bi-nucleate cell; means of genetic transformation of plants.

Provirus: Virus integrated into a host cell's DNA.

Purine: A base present in nucleic acids [eg adenine & guanine in DNA & RNA].

Pyrimidine: Another base present in nucleic acids [eg cytosine, thymine in DNA; uracil replaces thymine in RNA].

Recombinant DNA: DNA synthesised by splicing selected DNA sequences excised from different organisms. [Also described as 'Genetic modification'; 'Genetic engineering'].

Regeneration: Development of whole organisms from cell cultures.

Reproductive clone: A new living organism indentical to a single parent (e.g. Dolly, the sheep)

Restriction enzyme: An enzyme that selectively, or at random, cuts and excises a sequence of a DNA molecule.

Retrovirus: Uses an enzyme reverse transcriptase to copy its RNA genome into DNA that integrates into a host cell genome.

Retroviral vector: Retrovirus used to transfer foreign DNA into animal cells; replaces part of viral genome with host genome.

Ribosome: Phytoplasmic particles of RNA plus protein. Sites where proteins are synthesised as directed by mRNA code.

RNA - Ribonucleic acid: A polymer of ribose, phosphate, purine and pyrimidine bases that transmits from DNA instructions for specific protein synthesis.

Serological typing: Antibody – antigen reactions by which pathogenic bacteria may be identified.

Somatic cell: Vegetative, non-reproductive cells as distinct from reproductive germ cells [Gk σοματικoζ = 'of the body'].

Somaclonal variation: During culture of somatic cells, DNA may reconstruct to form cells different from the parent. Progeny cells designated somaclonal variants; are sources of genetic variation.

Somatic hybridisation: Hybrids created by fusion of somatic cells as distinct from fusion of sexual gametes.

Spore: Reproductive body consisting of relatively few cells.

Sporophyte: A spore-bearing plant or organism.

Stigma: Female organ of a flowering plant: Distal end of the style from which deposited pollen passes to the ovule.

Teratogens: Agents that induce congenital malformations.

Therapeutic clone: A stem cell to be proliferated to replace or repair defective organ cells.

Thermophile: Microorganism that tolerates and proliferates at relatively high temperatures [ca 45°C]

Thymine: Pyrimidine nucleic acid base that pairs with adenine in DNA.

Ti plasmid: Plasmid of *A. tumefaciens* that transfers genes that induce tumors in plants. Disarmed form lacking tumor-inducing genes used to introduce foreign genes into plant cells.

Tissue culture: In vitro propagation of cells from animal or plant tissue

Totipotent: Cells capable of self-differentiation and development into whole organism or embryo.

Transcription: Synthesis of RNA from a DNA template. Genetic code in DNA puts bases in RNA into prescribed order.

Transfer RNA [tRNA]: Carries amino acids to ribosome where anticodon reads codon on mRNA and places amino acids in prescribed sequence.

Transposable elements: DNA pieces that move from one chromosome site to another or between chromosomes.

Transposon: Short mobile segments of DNA that insert themselves into different chromosome sites to cause mutations.

Uracil: Pyrimidine nucleic acid base that pairs with adenine in RNA.

Vaccine: Dead or attenuated pathogenic microorganism or virus which carries antigens. When injected into live animal stimulates production of antibodies to counteract the antigens and subsequent infection by the original pathogen.

Vector: A vehicle by which DNA is transferred between cells.

Virus: Smallest known organism which proliferate by infecting living cells to usurp their synthetic and reproductive processes.

Zygote: A fertilized egg.

References

Agriculture Canada [1989], *Growing Together*, Agriculture Canada Publications 5269/E

Altieri, M A [1983], *Agro-ecology: The Scientific Basis of Alternative Agriculture*, Westview Press, Boulder, Colorado.

Ames, B N [1983], 'Dietary Carcinogens and Anti-carcinogens', *Science* 221[4617], 1256–1264.

Ames, B N & Gold L S [1983], 'Environmental Pollution and Cancer', *Science and the Law* 25 July, 1–33.

Amos, W F [1997], 'Waste water in the Food Industries', *Food Science Technology Today* 11[2], 96–104.

Arizpe, L *et al* [1992], 'Population and Natural Resource Use', *In: An Agenda of Science for Environment and Eevelopment in Twenty-first Century*, Cambridge University Press.

Armstrong, K [1994], *A History of God*, Ballantyne Books, New York.

Armstrong, K [2005], A Short History of Myth, A Knopf Press, Canada.

Barlow, W [1997] , Prince Philip Lecture, Royal Society of Arts, London.

Baum, W C [1986], 'Partners Against Hunger: Consultative Group on International Research', World Bank, Washington

Beaton, G H [1991], Human Nutrient Requirements, Food, Nutrition and Agriculture, FAO, Rome.

Bindoff, S T [1950], *Tudor England*, Penguin Books, UK.

Blaxter, K [1986], *People, Food and Resources*, Cambridge University Press, UK.

Blaxter, K [1992], 'From hunting and gathering to agriculture', *In: For Better Nutrition in the Twenty-first Century*, Ed: Leathwood P, Horseberger M, James W P J.

Boeringer, R [1980], *Alternative Methods of Agriculture*, Elsevier Science Pub, Amsterdam.

Bouquet, A C [1962], *Comparative Religion*, Penguin Books, UK.

Bourne, M C [1977], *Post-harvest Losses*, Cornell University Press, Ithaca, USA.

Brandt, W [1980], *North-South: A Programme for Survival*, M I T Press, Cambridge, Mass, USA.

Brandt, W [1983], *Common Crisis*, Pan Books, London.

Brown, H [1954], *The Challenge of Man's Future*, Viking Press New York.

Brown, L R [1987], *Sustaining World Agriculture*, Worldwatch Institute, Washington DC.

Brown, L R & Wolf, A [1984], 'Soil Erosion', Paper No 60 Worldwatch Institute, Washington DC.

Brundtland, G H [1987a], *Our Common Future*, Oxford University Press, Oxford.

Brundtland, G H [1987b], *Food 2000: Global Policies for Sustainable Agriculture*, Zed Books, London.

Bunting, A H [1987], *Agricultural Environments: Characterisation & Mapping*, CAB International, Oxford, UK.

CAEFMS [1990], Agriculture and the Environment: Economic dimensions of sustainable agriculture. CAEFMS Publication No 1., University of Guelph, Canada.

Carson, Rachel [1963], *The Silent Spring*, Houghton, Mifflin, Boston.

Chandler, R F [1982], *Renewal of the CGIAR: Sustainable Agriculture for Food Security*, CGIAR, World Bank, Washington.

Clutton-Brock, J [1987], *A Natural History of Domesticated Mammals*, Museum of Natural History, London.

CGIAR/TAC [1989], *Sustainable Agricultural Production: Implications for International Agricultural Research*, TAC Secretariat, FAO, Rome.

CGIAR [2002], 'CGIAR Report for 2002', CGIAR Secretariat, World Bank, Washington.

Cohen, A [2003], While Canada Slept, McLelland & Stewart, Toronto.

Cooper, B M [1971], *Writing Technical reports*, Pelican-Penguin Books, UK.

Cowell, F R [1948], *Cicero and the Roman Republic*, Penguin Books, UK.

Critchfield, R [1981], *Villages*, Anchor Press, Doubleday, New York.

Crosland, M P [1978], Historical Studies in the Language of Chemistry, Dover Publications, New York.

Davis, P J *et al* [2003], Synthesis of novel starches, *In: Planta,* Starch/Stärke **55**, 107–120.

De Burgh, W G [1947], *The Legacy of the Ancient World*, Macdonald and Evans, London.

Descartes, R [1966], Discourse on method, *In: The Philosophers of Science*, Washington Square Press, New York.

Dekker, J [1991], Pesticides, IN: ICSU-CASAFA 1991.

Demek, T *et al* [2000], 'Endosperm Starch in Canadian Wheat Cultivars', *Cereal Chemistry* **76**, 694–698.

Dixon, B [1973], *What is Science for?*, Penguin Books UK.

Doyle, J J & Persley, G J [1996], 'Enabling safe use of biotechnology', World Bank, Washington DC.

EEC [1991], Council Regulation (EEC) No 2092/91, 'On Organic Production of agricultural products', European Commission, Brussels.

EEC [2002], 'Organic farming in the EU: Facts and figures', European Commission, Brussels.

ECDGA [2000], 'Organic farming: Guide to Community rules', European Commission: Directorate-General for Agriculture, Brussels.

Edwards, C A, Lal, R *et al* [1990], 'Sustainable Agricultural Systems', Soil & Water Conservation Society. Ankeny, IA, USA.

Eurobarometer [2003], 'Europeans and Biotechnology in 2002', E C Directorate for Research, Brussels.

Evans, E [1993], 'Biotechnology: Opportunities for Control of Insects and Vectors', Rockefeller Foundation, New York.

Evenson, R E & Gollin, D [2003], 'Assessing the impact of the Green Revolution: 1960 to 2000', *Science* **300**, 758–762.

Farrar, M T & Milton, J P [1972], *The Careless Technology*, Natural History Press, Garden City, New Jersey.

F A O [1983], 'Post-harvest Loss in Quality of Food Grains', FAO, Rome.

F A O [1984], 'Land, Food and People', FAO, Rome.

FAO [2000], 'The State of Food Insecurity Across the World', FAO, Rome.

Fischer, L [1954], *Gandhi: His Life and Message for the World*, Mentor Books: the New American Library, New York.

Flannery, T [2006], *The Weather Makers: the Future Impact of Climate Change*, Allen Lane, London.

Fowler, H W [1926], *A Dictionary of Modern English Usage*, Oxford University Press, Oxford, UK.

Fowler, H W [1930], *The King's English*, Oxford University Press, Oxford, UK.

Fox, J & Brown, D [1988], *The Struggle for Acountability*, M I T Press, Cambridge, Mass.

Frascati [1981], *Frascati Manual: The Measurement of Scientific and Technical Activities*, OECD, Paris.

Galbraith, J K [1965], The Underdeveloped Country, Massey Lectures, CBC Publications, Toronto.

Galbraith, J K [1973], Economics and the Public Purpose, Houghton Mifflin, Boston.

Giampetro, M & Bukkens S G F [1992], 'Sustainable Development: Scientific & Ethical Assessment', *Journal of Agricultural & Environmental Ethics* **5**[1]. 27–57.

Gibbon, E [1985], *The Decline and Fall of the Roman Empire – An Abridged Version*, Penguin Books, London.

Gilliand, B [1983], World Population and Food Supply, *Population & Development Reviews* 9[2], 203–211.

Gowers, Sir E [1954], 'The Complete Plain Words', H M Stationery Office, London.

Gribbin, J [2002], *Science: A History*, Penguin Books UK.

H & W Canada [1989], 'Assessing the Safety of Crops Developed through Wide-cross Hybridisation', Health and Welfare Canada, Ottawa.

H & W Canada [1990], 'Nutrition Recommendations', Report of a Review Committee: Health & Welfare, Canada.

H & W Canada [1992], 'Canada's Food Guide to Healthy Eating', Health & Welfare, Canada, Ottawa.

Harper, R F [1999], *The Code of Hammurabi, King of Babylon*, The Law Book Exchange, N Jersey.

Harwood R R [1990], 'A history of sustainable agriculture', *In: Sustainable Agricultural Systems* Ed: Edwards C A St Lucie Press, USA.

Harrison R G [2003], 'Global Sustainable Development in the Twenty-first Century', TERI Silver Jubilee Conference, New Delhi.

Harris J M *et al* [2001], *A Survey of Sustainable Development*, Island Press, Washington DC.

Henderson [2004], 'How we may have wiped out two close relatives', *The Times of London:* 05 October.

Hengerveld, H [2006], 'CO$_2$/Climate Report, Science and Technology Branch', Environment Canada, Toronto.

Hengeveld, H [1995], 'Understanding Atmospheric Change', Environment Canada SOE Rpt 952.

Howe, R W [1967], Black Africa, New Africa Library.

Hulse, J H [1991], Global perspectives on sustainable development, *Canadian Journal Agricultural Economics* 39, 541–551.

Hulse, J H [1991a], Nature, composition and utilisation of grain legumes, *In: Uses of tropical grain legumes.* ICRISAT, Hyderabad, India.

Hulse, J H [1994], 'Nature and composition of food legumes', *In: Cool season food Legumes*, Ed. Muehlbauer F J & Kaiser W J: Kluwer Academic Publisher Netherlands.

Hulse, J H, [1995], *Science, Agriculture and Food Security*, NRC Research Press, Ottawa.

Hulse, J H [1996], Food technology: Internationally neglected, *Food Science and Technology Today* 10[4]. 194–197.

Hulse, J H [1998], 'Rural agroindustries and urban food security', Institute of: Ecotechnology and Agroindustries, M S Swaminathan Research Foundation, Chennai.

Hulse, J H [1999] , 'Urban food security: A crisis for food systems analysts', *Food Science & Technology Today* 13[3], 123–128.

Hulse, J H [2002], 'Ethical issues in biotechnologies and international trade', *Journal of Chem Technol Biotechnology 77*, 607–615.

Hulse, J H [2003], 'Biotechnologies: Past history, present state, future prospects', *Trends in Food Science & Technology* 15, 3–18.

Hulse, J H [2004}, Integrated food systems, *Journal of Food Science & Technology* 18[2], 44–45.

Hulse, J H & Escott, V J[1986], Drought: Inevitable and unpredictable, *Interdisciplinary Science Reviews* 11[4], 346–358.

Hulse, J H & Laing E M [1974], Nutritive value of the triticale protein, IDRC 021e, Ottawa.

Hulse, J H, Laing, E M Pearson, O E [1980], *Sorghum and the Millets: Composition and Nutritive Value*, Academic Press, London.

Hutchinson, J [1974], *Evolutionary Studies in World Crops*, Cambridge University Press, UK.

IBSRAM [1991], Evaluation for sustainable land management in the developing world, International Board for Soil Research & Management, Bangkok, Thailand.

ICSU-ASCEND [1992], *An Agenda of Science for Environment and Development for the Twenty-first Century*, Cambridge University Press.

ICSU-CASAFA [1991], *Sustainable Agriculture and Food Security*, ICSU, Paris.

IDRC [1981], *A Decade of Learning*, International Development Research Centre, Ottawa.

IFBC [1990], Safety of foods produced by genetic modifications, *Regulatory Toxicology & Pharmacology* 12[3], S1 –S190.

IFPRI [2002], Sustainable Food Security for all by 2020, International Food Policy Research Inst, Washington DC.

IFST [1996], Guide to Food Biotechnology, Institute of Food Science and Technology [UK].

IFT [2000], IFT Expert report on biotechnology and foods, *Food Technology* 54[8].

IFT [2001], Detection of GM food products, *I F T Journal of Food Sciences* (USA) 66[5].

ISRIC [1991], World map of the status of human-induced soil degradation, International Soil Reference & Information Centre, Wageningen, Netherlands.

IUNS/CABI [1983], Recommended dietary intakes around the world, *Nutrition Abs Rev* 53[11], 939–1117.

Jain, H K [1983], 'Agriculture and Environment', *Indian Farming*, June 1983, 5–19.

James, C [2005], 'Global status of Commercialized Biotech/GM Crops 2005', ISAAA, Ithaca, New York.

Jogdand S N [1996], *Environmental Biotechnology: Industrial Pollution Management*, Himalaya Publications House, Mumbai

Jogdand S N [2003], *Operational Maintenance of Wastewater Treatment Plant*, Book Enclave, Jaipur.

Jones, Joseph M [1965], *The United Nations at Work*, Pergamon Press, London.

Kates, R W [1994], 'Sustaining life on earth', *Scientific American, 271[4]*, 92–99.

Kennedy, Paul [1988], *Paris 1919*, Random House, New York.

Kidd, Charles V [1992], The evolution of sustainability, *Journal of Agricultural & Environmental Ethics 5*[1], 1–26.

League of Nations {1936–1937], 'The Relation of Nutrition to Health, Agriculture and Economic Policy', Report of Committee on Nutrition, Geneva.

Leick, G [2002], *Mesopotamia*, Penguin Books, UK.

Lewis, BW [1993], *The Arabs in Hisory*, Oxford University Press, UK.

Livy: Titus Livius [1960], *The Early History of Rome*, Penguin Books, UK.

MacIntyre, A [1966], *A Short History of Ethics*, Collier Books, New York.

McEvedy, C [1976], *The Penguin Atlas of Ancient History*, Penguin Books, UK.

McMillan, Margaret [2003], *Paris 1919*, Random House, New York.

Meadows, D M *et al* [1972], *The Limits to Growth*, Universe Books, New York.

Meadows, D L [1977], *Alternatives to Growth*, Bullinger Publishing, Boston.

Mees, C E K 1947, *The Path of Science*, J Wiley, New York.

Mookerji, R K [1961], *Glimpses of ancient India*, Bhavan's Book University, Bombay.

NABC [1989], Biotechnology and Sustainable Agriculture, National Agricultural Biotechnology Council, Boyce-Thompson Inst, Ithaca, New York.

N A S [1978], 'Post-harvest food losses in developing countries', National Academy of Sciences, Washington DC.

Nowell-Smith, P H [1959], *Ethics*, Pelican-Penguin Books, London.

NRC/NAS [1977], *World Food and Nutrition Study*, National Academic Press, Washington DC.

NRC/NAS [1989], *Alternative Agriculture*, National Academic Press , Washington DC.

Nye, R B & Morpurgo J E [1955], *A History of the United States*, Penguin Books, UK.

P C F F F [2002], Farming & Food: A Sustainable Future, UK Ministry of Environment, Food & Rural Affairs.

Pearson, L B [1969], *Partners in Development*, Praeger Publishers, New York.

Plumb, J H [1963], *England in the Eighteenth Century*, Penguin Books, UK.

Plumptre, A F W [1977], *Three Decades of Decision*, McClelland and Stewart, Toronto.

Prabhu, R K [1960], *An Anthology of Modern Indian Eloquence*, Bharatiya Vidya Bhavan, Bombay.

Price, T G [1989], *The Chemistry of Prehistoric Human Bone*, Cambridge University Press, UK.

Quiller-Couch, Sir A [1916,1938], *On the Art of Writing*, Cambridge University Press, Cambridge, UK.

Rama [1950], If you must write – Peter Garnett, London.

Reganold, J P & Papendick R I[1990], 'Sustainable agriculture', *Scientific American* **262**[2], 112–120.

Reutlinger, S[1987], Food security, trade and aid. *In: Food Policy, Ed.* J P Gittinger *et al*. John Hopkins University Press, Baltimore.

Revelle, R [1976], 'Resources available for agriculture', *Scientific American* **253**[9], 165–178.

Rodale, R [1983], *Breaking New Ground, The Futurist* Vol 1.

Roy Soc [2004], Nanosciences and Nanotechnologies. Royal Society Policy Document 19/04; Royal Society, London.

Rose, S [1970], *The Chemistry of Life*, Penguin Books, UK.

Rosegrant, M W *et al* [2002], Global Water Outlook to 2025, International Food Policy Research Inst, Washington DC.

Russell, B [1957], *History of Western Philosophy*, Allen & Unwin, London.

Russell, B [1967], *An enquiry into Meaning and Truth*, Pelican-Penguin Books, London.

Santorum, A & Gray J [1993], A general approach to assessment of public stocks. IN: *Food Security & Food Inventories in Developing Countries, Ed.* P Beck & B Digma CABI Press, Oxon.

Schumm, M K 1987, *Understanding Organic Chemistry*, Macmillan Pub Co. N Y.

Scientific American [1994] 'Sustaining life on earth', *Scientific American* **271**[4], 22–99.

Sen, A [1987], Poverty and entitlements. *In: Food Policy, Ed.* J P Gittinger *et al*, John Hopkins University Press, Baltimore.

Simmonds, N W [1976], *Evolution of Crop Plants*, Longmans, UK.

Stern, C & Sherwood E R [1966], *The Origin of Genetics*, Freeman & Co, San Francisco.

Stoskopf, N C [1993], *Plant breeding: Theory and practice*, Westview Press, Boulder, Colorado.

Swaminathan, M S [1999], Science in response to human basic needs, *Current Science* **77**[3], 341–353.

Swaminathan, M S [1999a], *A Century of Hope*, East-West Books, Madras.

Swaminathan, M S [2000], An Evergreen Revolution, *Biologist* **47**[2], 85–89.

Swaminathan, M S [2001], Achyara N G Ranga Memorial Lecture, M S Swaminathan Research Foundation, Chennai.

Swaminathan, M S [2002], *From Rio de Janeiro to Johannesburg*, East-West Books, Madras.

Swaminathan, M S & Sinha S K [1986], *Global Aspects of Food Production*, Tycooly International Press, Oxford, UK.

Tacitus C [1956], *Annals of Imperial Rome*, Penguin Classics, UK.

Thapar, R [1966], *A History of India*, Penguin Books, UK.

Thomas, C D *et al* [2004], *Nature* 427 (Jan 2004) 145–148.

Vepa, S S & Bahavani R V[2001], *Food Insecurity Atlas of Rural India*, M S Swaminathan Research Foundation, Chennai.

Verpmann, H-P [1973], Animal bone finds and economic archaeology, *World Archaeology* 4, 307–322.

Ward, Barbara [1961], *The Rich Nations and the Poor Nations*, Massey Lectures, C B C Publications, Toronto.

Ward, B & Dubos, R[1972], *Only One Earth*, Penguin Books, UK.

Weinberg, S [1994], Life in the Universe, *Scientific American* 271[4], 22–27.

Wells, H G [1942], *The Outlook for Homo Sapiens*, Readers Union: Secker & Warburg, London.

Wheatley, A D [1994], Effluent treatment in the food industry, *Food Science Tech Today* 8[1], 16–23.

Wheeler, A, Jones, A K [1989], *Cambridge Manuals in Archaeology*, Cambridge University Press UK.

WHO/FAO [1991], *Strategies for Assessing Safety of Foods Produced by Biotechnology*, World Health Organisation, Geneva.

Williams, P C [1974], 'Errors in protein testing', *Cereal Science Today* 19. 280–286.

Winegard, W [1987], For whose benefit? Report on Canada's Official Development Assistance, Government of Canada, Ottawa.

World Bank [1987], Social Indicators of Development 1987.

World Bank [1978–2006], World Development Reports (published annually).

Wortman, S & Cummings, R [1978], *To Feed this World*, John Hopkins University Press, Baltimore.

W R I [1997], Report of World Resources Institute, New York.

WRI [2004], Report of World Resources Institute, New York.

Zandstra, H G *et al* [1981], 'A Methodology for on-farm Cropping Systems, International Rice Research Institute Los Banos, Philippines.'

Index

I